Cambridge Lower Secondary

Complete Biology

Ann Fullick

Second Edition

WORKBOOK

Oxford excellence for Cambridge Lower Secondary

OXFORD

OXFORD
UNIVERSITY PRESS

Great Clarendon Street, Oxford, OX2 6DP, United Kingdom

Oxford University Press is a department of the University of Oxford. It furthers the University's objective of excellence in research, scholarship, and education by publishing worldwide. Oxford is a registered trade mark of Oxford University Press in the UK and in certain other countries

First published in 2021

British Library Cataloguing in Publication Data
Data available

978-1-38-201846-3

10

Paper used in the production of this book is a natural, recyclable product made from wood grown in sustainable forests.
The manufacturing process conforms to the environmental regulations of the country of origin.

Printed in India by Multivista Global Pvt. Ltd

Acknowledgements
The publishers would like to thank the following for permissions to use copyright material:

Cover: htu/Getty

Photos: p38: Ody_Stocker/Shutterstock; p38: Victorija Reuta/ Shutterstock.

Artwork by Aptara, Barking Dog Art, Erwin Haya, Integra, Q2A Media, and OUP.

Every effort has been made to contact copyright holders of material reproduced in this book. Any omissions will be rectified in subsequent printings if notice is given to the publisher.

This Student Workbook refers to the Cambridge Lower Secondary Science (0893) Syllabus published by Cambridge Assessment International Education.

The manufacturer's authorised representative in the EU for product safety is Oxford University Press España S.A. of El Parque Empresarial San Fernando de Henares, Avenida de Castilla, 2 – 28830 Madrid (www.oup.es/en or product.safety@oup.com). OUP España S.A. also acts as importer into Spain of products made by the manufacturer.

Introduction

Welcome to your **Cambridge Lower Secondary Complete Biology Workbook**.

This workbook accompanies the Student Book and includes one page of questions for every two pages of the Student Book. Each question page includes several types of question.

- Some questions ask you to choose words to complete sentences. These questions will help you to learn and remember key facts about the topic.
- Other questions ask you to identify statements as true or false, or put statements in the correct order. Some of these questions are testing your knowledge, others are asking you to apply what you know to a new situation.
- There are many questions that ask you to interpret data from investigations, or information from other sources. When you answer these questions, you will be practising important science skills, as well as preparing for the Cambridge Checkpoint test.
- Some pages include comprehension questions. They ask you to read some information, and then answer questions about it. Many of these questions will help you develop skills of evaluation.
- Most pages have an extension box. Some of these questions will help you to extend and develop your science skills. Many others go beyond Cambridge Lower Secondary Science. They include content equivalent to Cambridge IGCSE and O level. All the extension questions are designed to challenge you, and make you think hard. There aren't any spaces for your answers to these extension questions, so you'll need to work on a separate sheet of paper.

This workbook has other features to help you succeed in Cambridge Checkpoint and eventually Cambridge IGCSE:

- The glossary explains the meanings of important science words.
- The questions are excellent practice for the Cambridge Checkpoint test. They have been especially written by the author to provide lots of practice before your exam.
- The Extension practice questions show you what you are aiming for. Give them a try!

I wish you every success in science, and hope you enjoy the workbook.

Contents

Stage 9

1.1 What is life?

1. Draw lines to match each word to the correct definition.

Word
A sensitivity
B movement
C reproduction
D nutrition
E excretion
F growth
G respiration

Definition
1 the removal of waste products from a living thing
2 a process making the energy in food available to organisms.
3 detecting changes in the surroundings and responding to them
4 getting bigger, especially young organisms
5 moving all or part of the body
6 making more of the same type of organism
7 taking in or making food

2. Tick the statements that apply to **all** living things.
   - **a.** They sense their surroundings.
   - **b.** They can only store waste products for a short time.
   - **c.** They are able to make their own food.
   - **d.** They release energy using chemical reactions.
   - **e.** They are able to produce offspring.
   - **f.** They are capable of movement.
   - **g.** They increase in size during their lifetime.
   - **h.** They must feed their offspring.

3. State three pieces of evidence that support the idea that trees are living things.

....

....

....

Extension

In 2021, scientists from the USA and from China landed spacecraft on the planet Mars. They hope to bring samples of the surface of Mars back to Earth and look for signs of living organisms. Life on other planets may be too small for us to see, or very different to life on Earth. Scientists look for signs of life in any specimens they get. For example, they look for evidence of respiration.

a. Suggest two pieces of evidence that might show respiration is occurring.

b. Suggest one other piece of evidence scientists might look for in a sample of space material to see if life is present.

1.2 Investigating living organisms: yeast

Thinking and working scientifically

1. Read the following paragraph about planning a scientific investigation. Fill the gaps with words from the box below. Each word may be used once, more than once, or not at all.

   controlled **same** **measure** **question** **effect** **change**

   To plan an investigation, we need to decide what to .... to answer a ...., and what to .... to show the .... of that change. Other variables that could affect the results need to be ...., which means they are kept the ....

2. A student placed yeast in four flasks of warm water. She added different nutrients to each flask. Then she placed a balloon over the neck of each flask. After 30 minutes she measured the diameter of each balloon.

Nutrient added	Diameter of balloon (cm)
glucose	9.5
lactose	2
sucrose	5
fructose	4

   a. State which variable the student changed ....

   b. State which variable the student measured ....

   c. List three variables that must be kept the same.

   ....

   ....

   ....

   d. Name the gas that fills the balloons ....

   e. Name the reaction that produces the gas. ....

   f. Name the nutrient that lets yeast respire fastest ....

   g. We control the variables in an investigation to try and make it a fair test. Describe what we mean by a fair test.

   ....

   ....

Extension

Several different groups carry out the same investigation using the same apparatus, but their results are not exactly the same.

a. Suggest two reasons why they did not get the same results.

b. Explain the problems of carrying out a fair test using living organisms.

1.3 Classification and species

1. Choose the best words from the box below to complete the following sentences.

infertile **world** **Latin** **hybrids** **characteristics** **species**

A species is a group of plants or animals that share many of their .... Members of the same .... can breed with each other to produce offspring that are also able to breed. Each species has been given a two-part .... name that is used all over the .... Members of different species do not usually breed with each other. Those that do usually have .... offspring called ....

2. The two-part Latin name for a zebra is *Equus zebra*. Decide whether the animals below are from the same species, similar species or very different species from the zebra.
   a. Write their names in the correct columns in the table.

Equus ferus *Tetracerus quadricornis* *Syncerus caffer* *Equus africanus*

Same species	Related species	Very different species

   b. Explain your choices.

....

....

3. Read statements a–e below. Write T after the ones that are true and F after the ones that are false. Write the correct version of false statements on the lines below.
   a. Similar species share the same two-part Latin name. ....
   b. Members of the same species should produce fertile offspring. ....
   c. Members of the same species always look similar. ....
   d. A hybrid is born when animals from different species breed. ....
   e. Members of different species usually have fertile offspring if they breed. ....

....

....

....

Extension

A dzo is a cross between a yak and domestic cow. It is much stronger than either of its parents and grows faster.

a. Suggest why dzos have not been bred to form large herds, despite being more useful than both cows and yaks.
b. Yaks and dzos look quite similar. Suggest how scientists could distinguish between them.

1.4 Classifying invertebrate animals

1. Choose the best words from the box below to complete the following sentences.

groups characteristics differences classification species backbones

Scientists use similarities and .... to put living things into .... This is .... The major groups of animals are vertebrates, which have .... and invertebrates, which do not. These are subdivided into smaller and smaller sub-groups with different .... The smallest sub-group is a ....

2. The diagram shows one animal from each invertebrate group. Label the diagram. State the name of each invertebrate group and describe one of the main characteristics that animals in the group share.

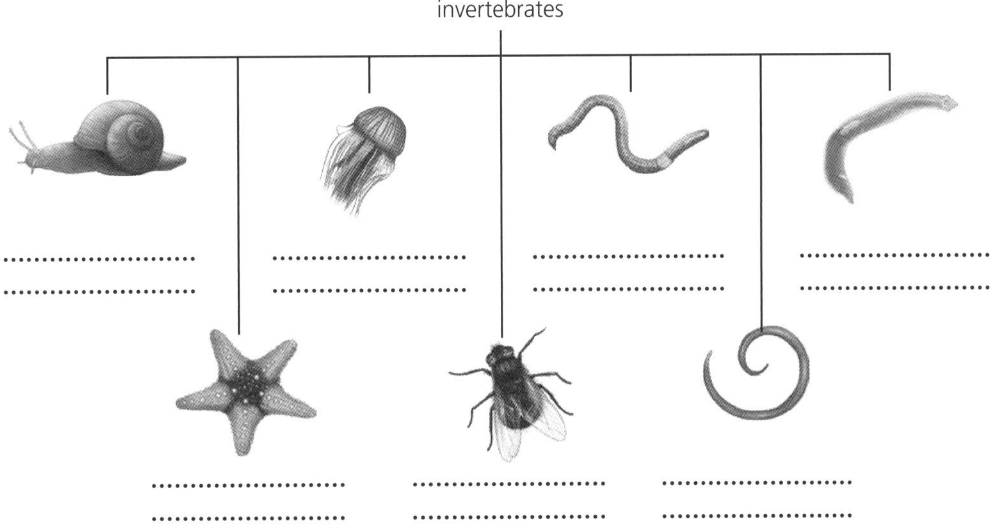

3. The diagram below shows four groups of arthropods. Label the diagram. State the name of each invertebrate group and describe one of the main characteristics that animals in the group share.

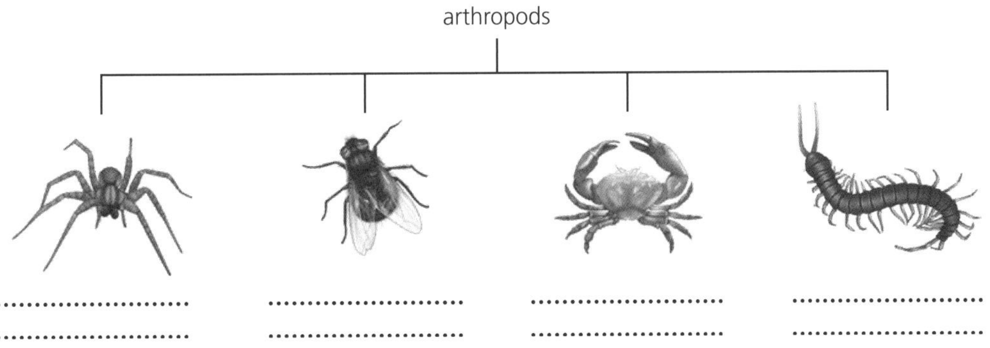

Extension

a. The drawing on the right shows a type of invertebrate. State what type of invertebrate it is and explain how you identify it from other arthropod and invertebrate groups.

b. i. These diagrams are secondary sources. Explain what this means.

ii. Give one advantage and one disadvantage of using secondary sources instead of direct observation.

Thinking and working scientifically

1. Each of these snakes has hollow fangs. Their fangs drop down when they open their mouths.

A

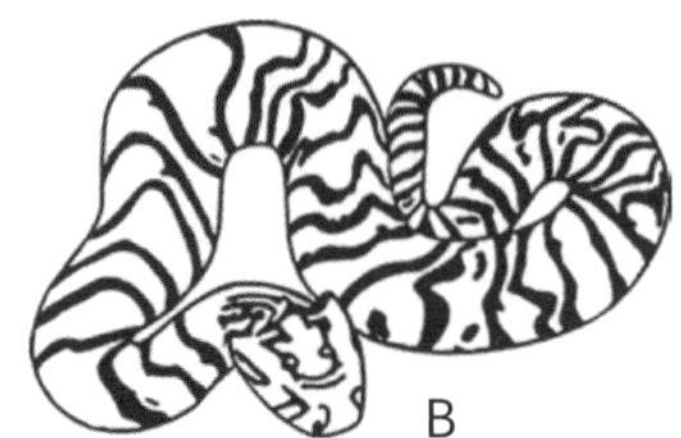
B

C

a. Use the following short key to identify which snake is *Bitis arietans.* ....

1	Hollow fangs drop down when it opens its mouth	YES: a type of venomous viper see Q2 NO: Not a venomous viper
2	Has a heart-shaped head	YES: a type of African adder see Q3 NO: Not a type of African adder
3	Has 18–22 V-shaped stripes	YES: *Bitis arietans (puff adder)* NO: other *Bitis* species

b. List the features you used to identify the snake.

....

....

c. State what you can say about the other two snakes based in the information in the key.

....

....

Extension

Discuss the advantages and problems of using a key to identify organisms based only on drawings, compared with using photographs or the actual living organisms.

1.6 Classifying vertebrates

1. The table below shows one animal from each of the five vertebrate groups.

   a. State the key characteristic of all vertebrates.

   ....

   b. Complete the table, stating the name of each group of vertebrates, and describing **three** key characteristics which help identify members of each group.

2. The echidna is a small vertebrate covered with hair and spines. Three weeks after mating, females produce an egg. This is kept in a pouch. After 10 days the egg hatches and the baby echidna feeds on its mother's milk.

   a. Which vertebrate group do echidna belong to? ....

   b. Describe two features this vertebrate group shares with echidnas.

   ....

   c. Describe one feature that makes echidnas different from the rest of the group.

   ....

Extension

The drawing shows an artist's reconstruction of an extinct vertebrate called *Archeopteryx*. It had claws, feathered wings, a long bony tail and a beak full of teeth.

a. Give one reason why *Archaeopteryx* is difficult to classify.

b. Which two vertebrate groups does *Archaeopteryx* share features with?

1.7 Classifying plants

1. a. Complete this table to show the features the plants in each group share.

Type of plant	Liverworts and mosses	Ferns	Conifers	Flowering plants
Roots and veins?				
Spores or seeds?				
Cones, flowers, or neither?				

b. Use your answers to **1. a.** to complete this key:

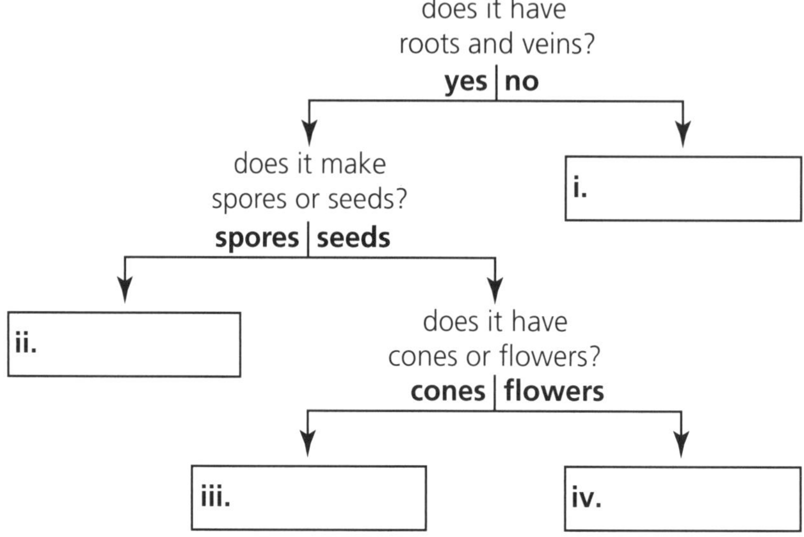

2. Identify the groups to which plants A–D belong. State which features you used to identify them.

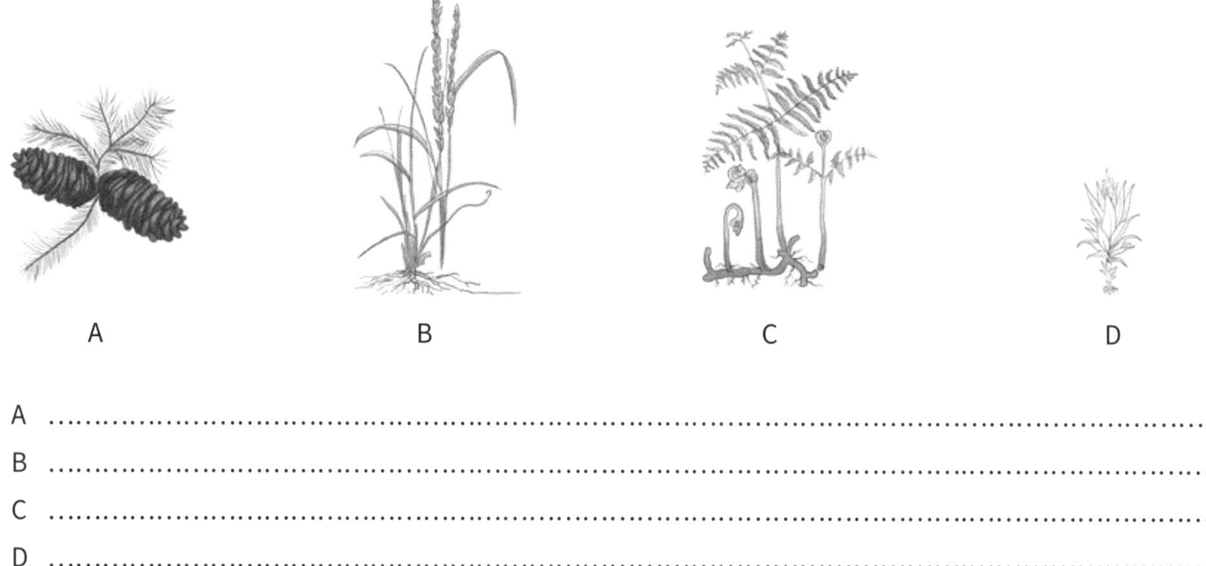

A ....

B ....

C ....

D ....

1.8 Making your own identification keys

Thinking and working scientifically

1.

The animals in this diagram are all birds.

Make a key to identify all six birds clearly. Produce your key either as a table or as a branching diagram.

1.9 Are viruses living?

Science in context

1. Complete the following statements about viruses. Use the words from the list below. You may use them once, more than once, or not at all.

living organisms diseases everywhere parasites invade bacteria virus-making factories

Viruses are found .... They are very small – even smaller than .... Viruses .... the bodies of .... and turn them into .... As a result, viruses cause many .... in other organisms. They are sometimes described as perfect ....

2. a. State two structures found in all viruses.

   ....

   b. Give two reasons why scientists find viruses hard to study.

   ....

   ....

   c. Name two human diseases caused by viruses.

   ....

3. Some biologists think viruses are living organisms. They are described as perfect parasites. Some think they are not living organisms.

   a. Explain what a parasite is.

   ....

   ....

   b. Complete this table. Suggest two reasons why viruses may be classified as living organisms, and two reasons why they may be classified as non-living.

Reasons to classify viruses as living organisms	Reasons not to classify viruses as living organisms

1.10 Moving classification forwards

Extension

Read this piece of information about how scientists are using new technology to help identify new species:

Many people, including scientists, are very worried about the loss of many species of animals and plants. Using traditional methods of collecting and observation to classify organisms is still important and widely used, but it takes a lot of time. Animals and plants of the same species may vary a lot. It is easy for scientists to think they have discovered a new species when it is just a different-looking member of an old species. In the same way, sometimes scientists think animals or plants belong to the same species when they are actually from different species. An example of this is the clouded leopards you studied in your student book.

To look after our living planet, we need to know what organisms live here. New technology means we can look at the genetic material of organisms and compare them to known species. One new method is called **DNA bar-coding**. Scientists look at small parts of the genetic material common to all known species of animals and plants, but which vary a lot. They can do this out in the field, wherever they are in the world, using small, hand-held devices.

1. a. Describe how organisms are classified using observation.

....

....

b. Suggest two reasons why this is still an important and useful way to classify new organisms.

....

....

2. a. Describe two advantages in using DNA bar-coding to help classify organisms in the field.

....

....

b. Suggest two disadvantages of using modern technologies to classify organisms.

....

....

3. Write a paragraph explaining why classification is important and how classification is changing.

....

....

....

....

....

....

2.1 The building blocks of life

1. Read the following paragraph. Complete the information using words from the box below. Each word may be used once, more than once, or not at all.

microscope	seven	respiration	80–100	bacteria
growth	cells	billions	excretion	20–30

All living things are made up of …………….. . Microorganisms such as protozoa, yeast and …………………… have only one cell, but large plants and animals have ……………………….. . Cells carry out the ……………….. processes of life: ………………., movement, ………………., reproduction, sensitivity, nutrition and ……………………… . Cells are very small – an average human cell is ……………. microns across. We need a special tool called a ………………………….. to see most cells.

2. This Venn diagram includes the features of light microscopes and two sorts of electron microscopes. TEM means transmission electron microscope. SEM means scanning electron microscope. SEMs are like TEMs but they take 3-D images of the outer surfaces of specimens. Write the letter of each statement in the correct part of the Venn diagram. For example, statement **a** is true for light microscopes and TEMs so we write **a** where those circles overlap.

   a. The specimen slice must be very thin.

   b. It magnifies the object up to 1000 times.

   c. It can magnify the object more than 1000 times.

   d. The images can be coloured artificially.

   e. It shows the surface of specimens and gives a 3-D image.

   f. Light passes through the specimen to make a magnified image.

   g. Used to look at organisms that are too small to see.

   h. Uses electrons to produce magnified images.

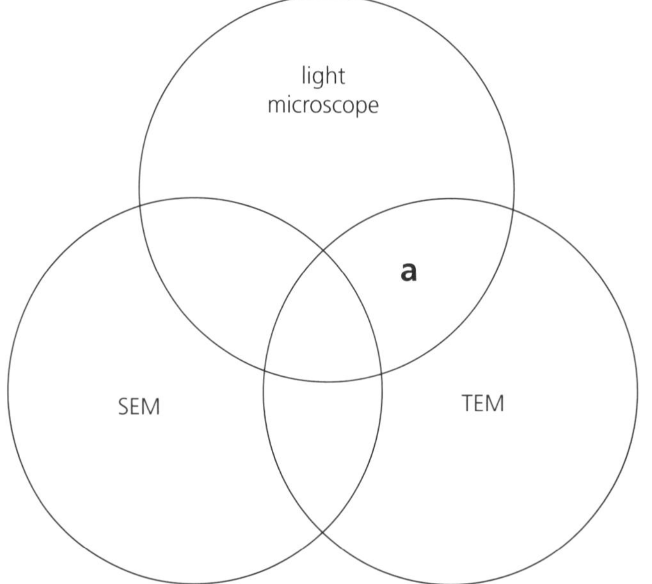

Extension

a. Give two advantages of using an electron microscope instead of a light microscope.

b. Give two disadvantages of using an electron microscope instead of a light microscope.

2.2 The cell story

Science in context

Answer the following questions about the discovery of the microscope. Answer in complete sentences.

1. State whether the first microscopes were made in the 17th century CE, the 18th century CE, or the 19th century CE.

   ....

2. Describe the first material seen under a microscope by Robert Hooke.

   ....

3. Anton van Leeuwenhoek was a tailor who used polished glass lenses to make a microscope in the 1670s and 1680s. Explain why he was so excited when he looked down his microscope at a drop of pond water.

   ....

   ....

   ....

4. Suggest two reasons why it took almost 200 years from the development of the first microscope before scientists all accepted the cell theory.

   ....

   ....

   ....

   ....

5. State the modern cell theory accepted by scientists everywhere

   ....

   ....

Extension

Write a letter from Anton van Leeuwenhoek to a friend describing the new microscope he has made and the exciting things he is seeing.

2.3 Animal and plant cells

1. Label the four parts of the animal cell shown below:

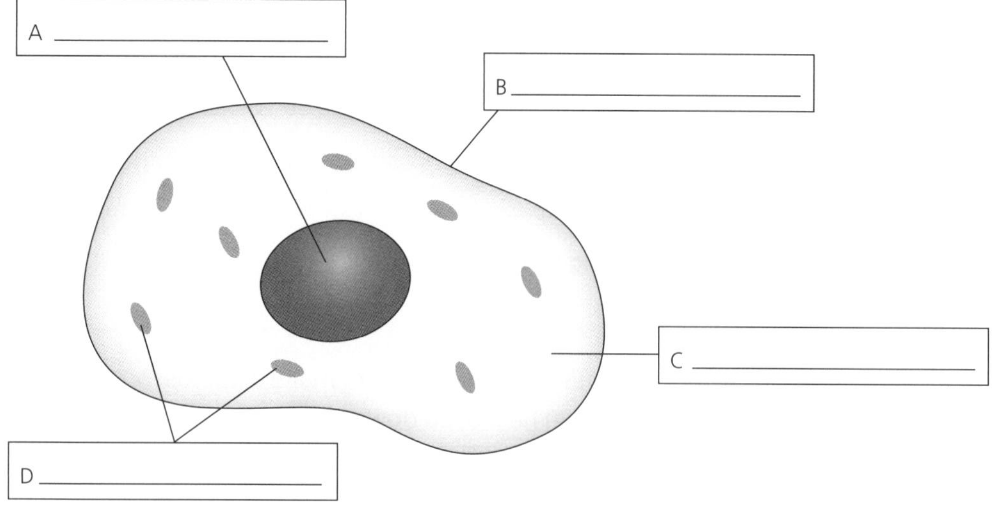

2. Label the seven parts of the plant cell shown below:

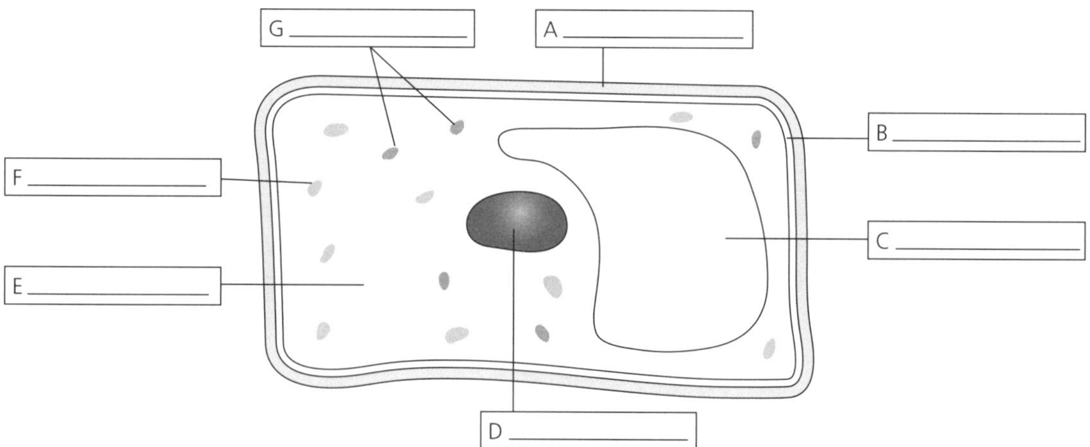

3. Draw lines to match each cell structure to its function in the cell:

Cell structure
A cell membrane
B vacuole
C chloroplast
D cell wall
E nucleus
F cytoplasm
G mitochondria

Function in the cell
1 tough outer layer giving plant cells shape and strength
2 where most chemical reactions take place in the cell
3 controls what enters and leaves the cell
4 contains the genes, controls the activities of the cell
5 membrane-lined space filled with cell sap in plant cells which pushes against the cell wall to keep plant cell firm
6 where respiration takes place, providing energy for all the chemical reactions of the cell
7 absorbs light so the plant cell can photosynthesise

Extension

Mitochondria are found in both animal and plant cells. Respiration takes place inside them. State whether you expect muscle cells or fat cells to have most mitochondria. Explain your choice.

2.4 Using a microscope

Thinking and working scientifically

1. Label this diagram of a light microscope.

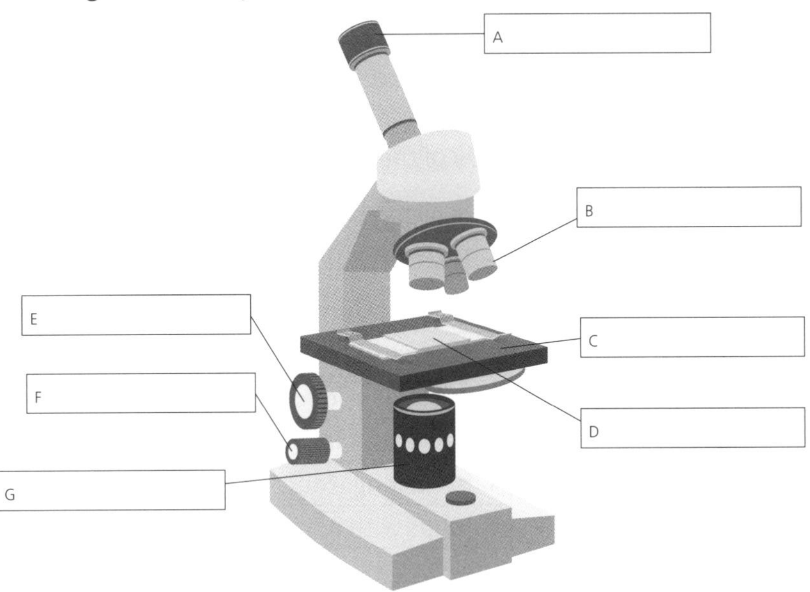

2. Number these statements on using a light microscope in the correct order to make a set of instructions for a student. Place your numbers in the boxes provided. Use pencil to begin with in case you make a mistake and want to erase it:
   - ☐ Choose the objective lens with the lowest magnification and move it into position.
   - ☐ Draw your observations using a pencil. Record the magnification you are using.
   - ☐ Move the stage of the microscope to its lowest position.
   - ☐ Once the object is in focus , turn the fine focus knob slightly to see if it makes the image even sharper.
   - ☐ Put the object you want to look at on the stage. Keep it in place using the clips.
   - ☐ To see your specimen in more detail, repeat these steps using an objective lens with a higher magnification.
   - ☐ Look through the eyepiece lens and turn the coarse focus knob slowly until the object comes into focus.

3. You know that

   total magnification = eyepiece lens magnification × objective lens magnification

   A student uses an eyepiece lens with a magnification of ×5. They use an objective lens with a magnification of ×4. State the total magnification of their specimen. Show your working.

   ....

Extension

A student looks at a plant specimen using a total magnification of ×100. Her eyepiece lens has a magnification of ×4. What is the magnification of the objective lens she is using? Show your working.

2.5 Specialised animal cells

1. Read the following paragraph and fill in the gaps with words from the box below. Each word may be used once, more than once, or not at all.

structure **specialised** **cells** **function** **multicellular** **respiration**

All living things are made up of .... Each cell uses energy from .... to stay alive. Many organisms are made up of many cells – they are .... These cells are not all the same. Many of them are .... cells. The .... is adapted to carry out a particular .... in the body of the organism.

2. Look at the different specialised animal cells listed here and the list of adaptations seen in these different specialised cell types. Match each adaptation to the correct specialised animal cell.

Cell type
A red blood cell
B muscle cell
C fat cell
D bone cell
E nerve cell (neurone)

Specialisation
1 contains fibres which can make themselves shorter
2 contains a lot of lipids which act as an energy store
3 contains haemoglobin to transport oxygen in the body
4 has a long extension to carry electrical messages μm
5 binds with mineral salts to make a rigid solid

3. More than half of the cells in your body are red blood cells. They are very small, only 7 μm in diameter, but they are very important because they carry oxygen around your body. Explain how the four features listed allow red blood cells to carry out their function in your body. Red blood cells:

- are filled with haemoglobin

  ....

  ....

- are small and flexible

  ....

  ....

- do not have a nucleus

  ....

  ....

- are biconcave

  ....

  ....

Extension

Describe the adaptations you see in ciliated cells and explain how these adaptations are linked to the functions of ciliated cells in the body.

2.6 Specialised plant cells

1. This diagram shows specialised plant cells called palisade cells.
   a. Name the process palisade cells are specialised to carry out. ....
   b. Describe three adaptations of palisade cells to this process. Explain how each feature makes the cell better adapted to their function.

Adaptation1...

Explanation...

Adaptation 2..

Explanation...

Adaptation 3 ....

Explanation...

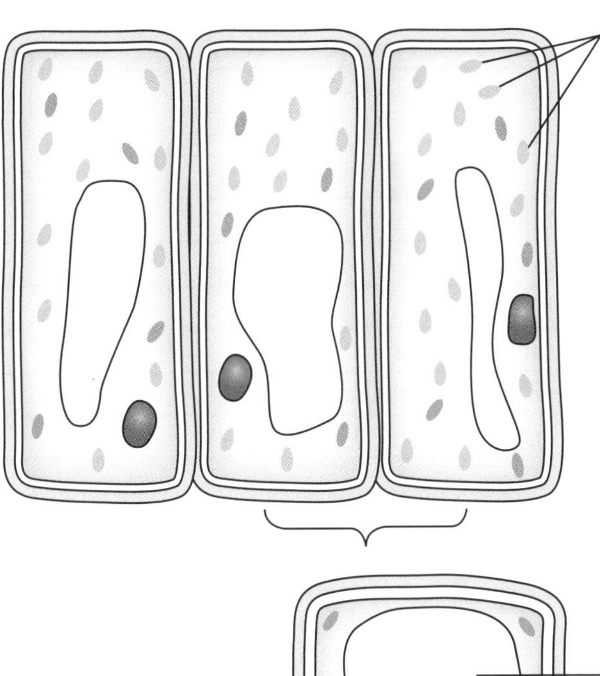

2. a. State the name of this type of specialised plant cell. ....

   ....

   b. Describe the main functions of these specialised cells.

   ....

   ....

   c. Label and annotate the diagram to describe and explain the three main adaptations of this type of plant cell to its functions.

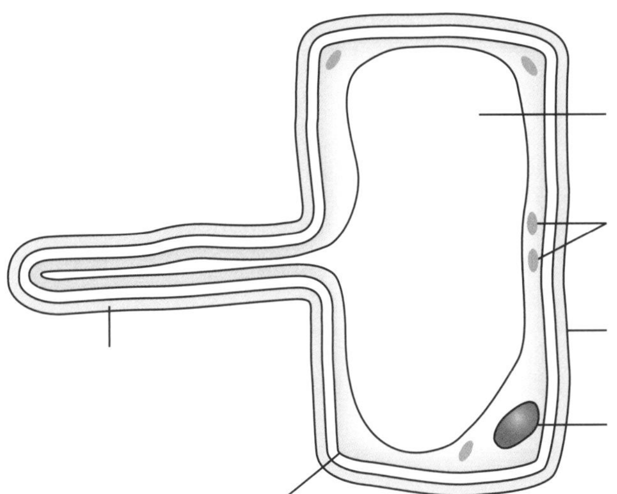

Extension

One important structure found in some plant cells is clearly visible in one of the specialised plant cells shown on this page and is not present in the other.

a. Suggest which important structures are present in only one of these specialised plant cells and state which type of cell it is in.
b. Explain why these structures are an important adaptation in one specialised type of plant cell but not in the other.

2.7 Modelling cells

Thinking and working scientifically

1. Complete the table below comparing plant and animal cells. Put a tick (✓) or a cross (x) in the column to show if a structure or process is present in animal cells and plant cells.

Feature	Animal cells	Plant cells
cell membrane		
cell wall		
nucleus		
large sap vacuole		
chloroplasts		
cytoplasm		
mitochondria		
carries out respiration		
carries out photosynthesis		

2. Scientists often use models. Read this paragraph about models and fill in the gaps, with words from the box below. Each word may be used once, more than once, or not at all.

small **remember** **misconceptions** **physical** **explain** **big**

Scientists often use .... models to help .... things that are very .... or very .... A bad model can leave you with .... but a good model will help you understand and .... the structure of a system.

3. Suggest four ways a good model of an animal cell and a plant cell will be the same, and three ways in which the models should differ if they are to avoid misconceptions.

....

....

....

....

....

....

....

....

....

....

....

....

....

....

2.8 Tissues and organs in animals

1. Many multicellular organisms have five levels of organisation. Match the levels of organisation given here with the correct description.

Level	Description
A cells	1 made up of several different tissues working together to carry out a particular function
B tissues	2 the building blocks of all living organisms
C organs	3 made up of many different organs and organ systems working together
D organ systems	4 made up of specialised cells of the same type working together to carry out a particular function
E whole organism	5 made up of several different organs working together to carry out a particular function

2. Label the main human organs shown on this diagram and add a brief description of their function.

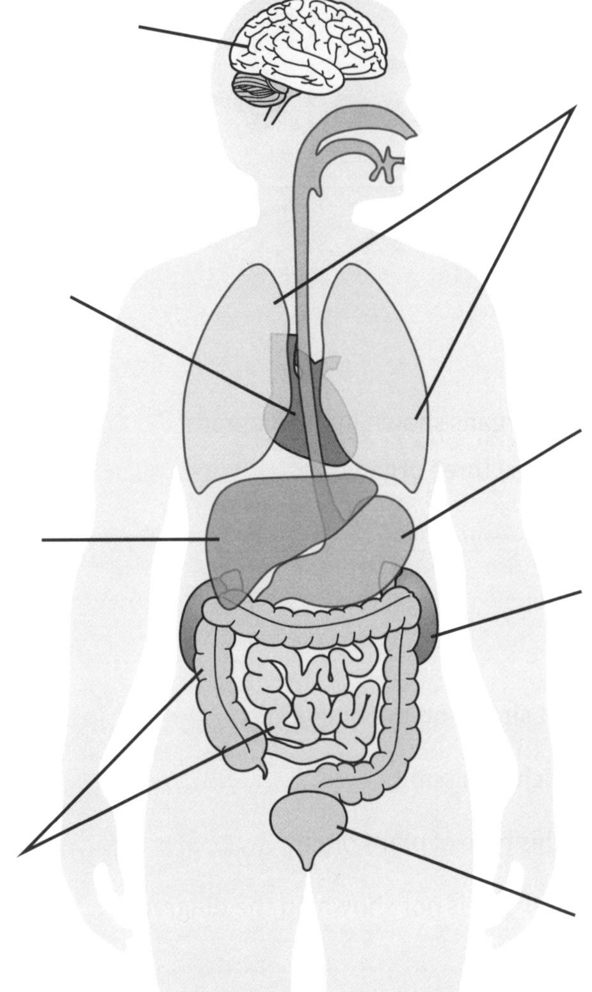

Extension

Choose one organ system in a mammal such as a human being. List as many of the organs in the organ system as you can. Describe the function of each organ. Then suggest some of the tissues which are involved in the different organs and explain how they help the organ system function.

2.9 Tissues and organs in plants

1. Read the following paragraph and fill in the gaps with words from the box below. Each word may be used once, more than once, or not at all.

root	cells	whole organism	shoot	organs
multicellular	photosynthetic	two	organisation	tissues

Many large .... organisms have five levels of .... .From the smallest to the largest these are .... → .... → .... → organ systems → .... Most plants have .... main organ systems – the .... or .... system above the ground and the .... system below the ground.

2.

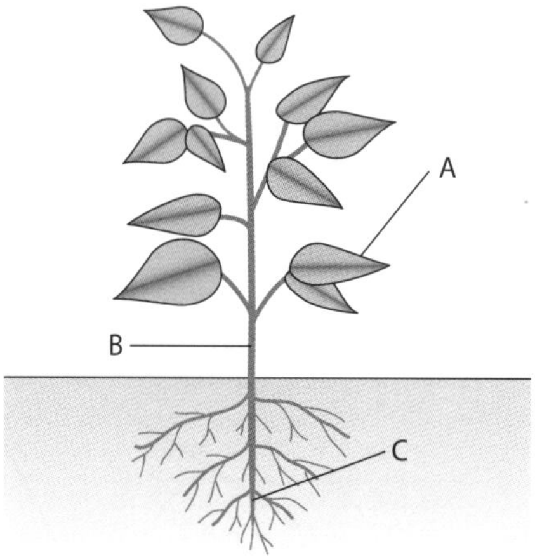

A

B

C

a. Label the three main plant organs shown in the diagram.

b. Describe the function of these three organs.

....

....

....

3. One type of plant organ is not shown on the diagram.

a. Name the plant organ which is missing. ....

b. Describe the function of this type of plant organ. ....

c. Suggest a reason why this organ is not shown on the diagram. ....

....

4. Identify each of these types of plant tissues:

a. Made up of dead cells forming hollow tubes to transport water and minerals. ....

b. Made up of living cells that transport dissolved sugar from the leaves to any plant cells that cannot make their own food. ....

c. Made of closely packed, brick-shaped cells that contain many chloroplasts. ....

3.1 Microorganisms

1. Read the following paragraph. Complete the information using words from the box below. Each word may be used once, more than once, or not at all.

fungi	microscope	single	bacteria	culture	viruses

Microorganisms are very important to life on Earth. Most microorganisms are .... cells. They include ...., ... and some .... We observe microorganisms using a .... or by growing a ....

2. Read the statements below and identify each type of microorganism:
   a. It is made up of a tangle of microscopic threads called hyphae and absorbs nutrients from its surroundings. It is a ....
   b. They are between 0.2 and 2.0 μm in diameter, about 1/10 the size of a human cell. They reproduce rapidly by splitting in two. They are ....
   c. They are the smallest microorganisms and are made of genetic material inside a protein coat. They are ....
   d. These round, single-celled microorganisms are bigger than bacteria. They reproduce by budding, a way of splitting in two. They are ....

3.

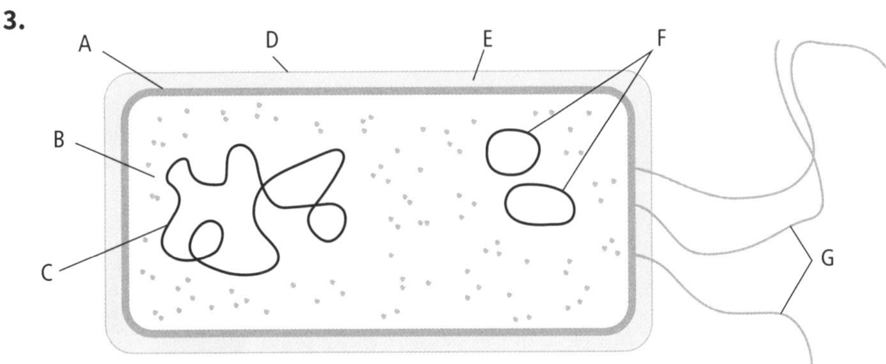

   a. State which letter represents the loop of genetic material seen in all bacterial cells. ....
   b. State which letter represents the small extra pieces of genetic information seen in some bacteria and name these structures. ....
   c. i. Give the label letter and name of two structures found in every bacterial cell.

      ....

      ii. Describe the function of the two structures you named in **c.i**.

      ....

      ....

Extension

Make a table to compare bacteria, fungi, and viruses.

3.2 Microorganisms are our friends

Extension

1. Read the following paragraph and fill in the gaps with words from the box below. Each word may be used once, more than once, or not at all.

respiration	yeast	dough	soft	carbon	dioxide	flour

Bread is made by baking ...., which is a mixture of .... and water. To make bread rise, ... can be added to the mixture. It uses nutrients from the flour for .... and releases bubbles of .... gas. The gas bubbles make the bread .... and spongy.

2. **a.** Bread is not the only food made using microorganisms. Name two more common foods made using microorganisms.

   ....

   b. Name one antibiotic and explain what these medicines do.

   ....

   ....

3. When yoghurt is made, bacteria feed on the sugars in the milk to produce a sharp, acidic taste. Two bottles of milk were kept in a warm place. One had extra bacteria added to it. The pH of each bottle of milk was measured at the start and after 4 hours. Look at the evidence in the bar chart.

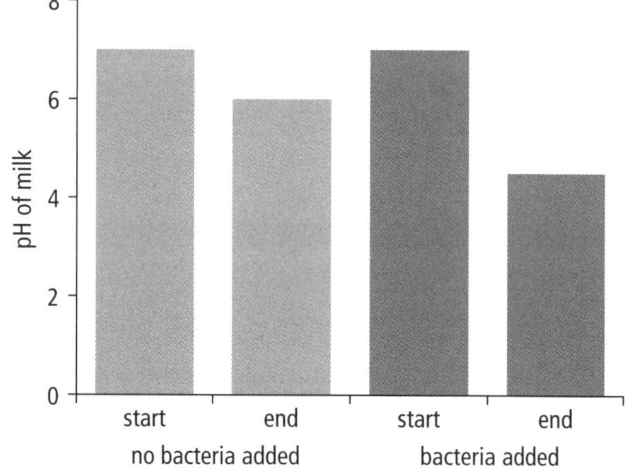

   a. Describe the effect of the extra bacteria on the milk mixture.

   ....

   ....

   b. Describe how the bacteria change the sugars in milk and explain how this affects the yoghurt that is made.

   ....

   ....

Extension

When yeast s are added to sugar solution they produce carbon dioxide. Nadia counted how many bubbles of gas were produced per minute by yeast/sugar solutions at different temperatures.

a. Display these results on a graph and draw a curve through the points.

b. Suggest reasons for the shape of the graph.

Temperature (°C)	Bubbles per minute
30	60
35	78
40	100
45	52
50	5

3.3 Microorganisms and disease

Thinking and working scientifically

1. Louis Pasteur developed a hypothesis that germs (microorganisms) caused disease. Number the boxes to put the following sentences in order to show how Pasteur developed and tested his hypothesis.

- ☐ Pasteur decides to grow anthrax germs, weaken them, and inject them into a sheep to protect it against the disease.
- ☐ Pasteur has a hypothesis that infectious diseases like anthrax are caused by germs passed from one person or animal to another.
- ☐ Three days after the animals were infected with anthrax, there was a public inspection. All the vaccinated animals were alive and healthy. All the unvaccinated animals were dead or dying of anthrax.
- ☐ Hippolyte Rossignol does not believe in germ theory and challenges Pasteur to prove his theory with a public test.
- ☐ A few weeks later all the animals are infected with live anthrax from infected animals.
- ☐ Pasteur predicts that the vaccinated animals will survive and the unvaccinated animals will develop anthrax and die.
- ☐ Pasteur had clearly demonstrated both the germ theory of disease and the value of vaccination.
- ☐ Pasteur gives a group of healthy sheep, cows and goats his vaccine. He selects a similar group of animals that do not get the vaccine.

Extension

Louis Pasteur used flasks like these to demonstrate that there are microorganisms in the air and they cause things to go bad. He needed this evidence before he could produce his germ theory of disease. Look at the statements below and draw lines to match what Pasteur does in this investigation with a stage of developing a scientific explanation.

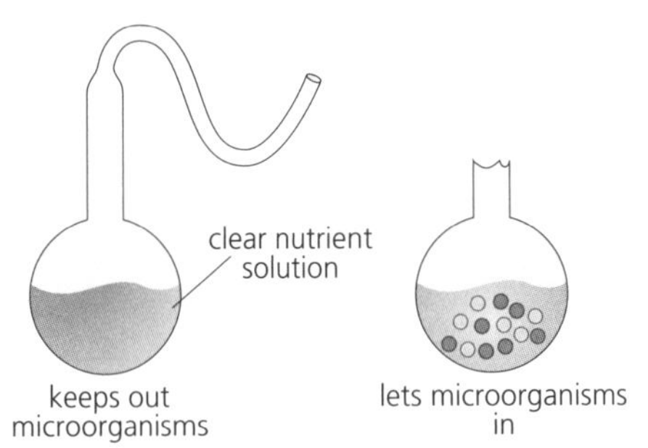

What Pasteur does
Says: When microorganisms land in liquids full of nutrients, they grow and reproduce and make the liquid cloudy.
Thinks: If I keep microorganisms out of a nutrient solution, it will not go cloudy because it will not go bad.
Places nutrient solutions in flasks with S-shaped necks and boils them to destroy any microorganisms present.
Notices that nutrient solutions do not go cloudy in flasks with S-shaped necks.
Breaks the neck of one of the flasks and observes that the nutrient solution begins to ferment and turn cloudy as microorganisms get in.

Stage of developing an explanation
make a prediction
suggest an explanation
review the evidence
collect extra evidence
test the explanation

Science in context

1. Complete this table to describe these four ways in which disease-causing microorganisms are spread from person to person.

How disease-causing microorganisms spread from person to person
Droplet infection
Contact
Contaminated food and drink
Break in the skin

2. a. Describe three ways of avoiding the spread of infectious diseases caused by droplet infection. Use your scientific understanding to explain how each method works.

   ....

   ....

   b. Describe two ways of avoiding the spread of infectious diseases caused by contact with other people and use your scientific understanding to explain how each method works.

   ....

   ....

Extension

The pathogen that causes typhoid is present in faeces. Suggest three ways to reduce the spread of this disease and explain the scientific understanding that supports each of them.

3.5 The decomposers

1. Read the following paragraph and fill in the gaps with words from the box below. Each word may be used once, more than once, or not at all.

bacteria	**mineral salts**	**herbivores**	**decomposers**	**nutrients**
water	**dead**	**fungi**	**plants**	

Plants take .... and .... from the soil and use them to help build new plant material. ....eat the plants. The organic waste from animal droppings and .... animals and plants is broken down by .... These organisms include .... and .... They use some of the .... and release others back into the soil to be taken up and used again by ....

2. People use decomposers to make compost.

   a. Describe decomposers.

   ....

   ....

   ....

   b. People use decomposers to make compost. Describe compost.

   ....

   ....

   c. Explain the importance of using decomposers to make compost.

   ....

   ....

   ....

Extension

Sewage is the term we use for human bodily waste.

a. Explain why sewage is a growing problem.

b. Describe how the decomposers can help solve the problems of sewage.

c. Explain why it is important to get rid of human sewage.

Thinking and working scientifically

1. In a scientific investigation, you make conclusions based on your results. You must evaluate your investigation to identify problems and see if you can suggest improvements.
   a. State what is meant by an anomalous result.

   ....

   ....

   b. Describe the three main stages of a good evaluation.

   ....

   ....

   c. Explain the difference between random and systematic errors in your results.

   ....

   ....

2. A group of students is investigating the effect of temperature on the rate of decomposition of some tomatoes. They put four tomatoes on a plate at room temperature. They put another four tomatoes in a sealed plastic bag at room temperature. They repeated this, putting the tomatoes in a fridge. They observed the rate at which the tomatoes decomposed in the different settings by estimating the percentage of the tomatoes covered in mould. The results are shown in the table below:

Conditions	Percentage of tomatoes covered in mould				
	Day 0	**Day 5**	**Day 10**	**Day 15**	**Day 20**
room, open plate	0	10	30	50	90
room, plastic bag	0	5	15	25	75
fridge, open plate	0	0	5	15	40
fridge, plastic bag	0	0	0	5	10

   a. Explain the main observations made by the students.

   ....

   ....

   b. Evaluate the method used by the students. Identify any errors as fully as you can.

   ....

   ....

   ....

   c. Suggest two ways in which this investigation could be improved.

   ....

   ....

Extension

Draw a graph of the results obtained by the students in their tomato experiment.

3.7 Food chains, food webs and decomposers

1. Draw lines to match each word to the correct definition.

Word
A predator
B prey
C consumer
D producer
E herbivore
F carnivore
G scavenger

Definition
1 eats other living things to obtain nutrients
2 hunts and eats other animals
3 is hunted and eaten by other animals
4 only eats plants
5 makes its own food
6 only eats animals that are already dead
7 only eats animals

2. Look at this food chain:

maize → mice → snakes → mongooses

a. Name the producer in this food chain. ....

b. Name a consumer in this food chain which is also a herbivore. ....

c. State a consumer which is both predator and prey. ....

d. Explain why all the arrows in the food chain point away from the producer. ....

....

....

e. Add the decomposers to the diagram and describe what they do. ....

....

3. This diagram shows part of a desert food web.

a. Define the term 'food web'.

....

....

....

b. Explain the advantages of using a food web instead of a food chain.

....

....

....

c. Name two herbivores and two carnivores in this food web.

....

....

d. Two important types of organisms are missing from this food web. State what they are and explain why each is so important.

....

....

4.1 The physical environment

1. Read the following paragraph. Complete the information using words from the box below. Each word may be used once, more than once, or not at all.

temperature	**water**	**fertile**	**abiotic**	**rocks**	**cold**

Non-living factors in the environment are known as .... factors. They include the .... or soil, the .... and the amount of .... available. Organisms growing in rocky or hot places will be very different from those growing in .... soil or a .... climate. In all environments, organisms need .... to live.

2. **a.** State two ways in which water is used in living organisms.

....

....

b. About 60% of the human body is water. The graph below shows the percentage of water in different human organs.

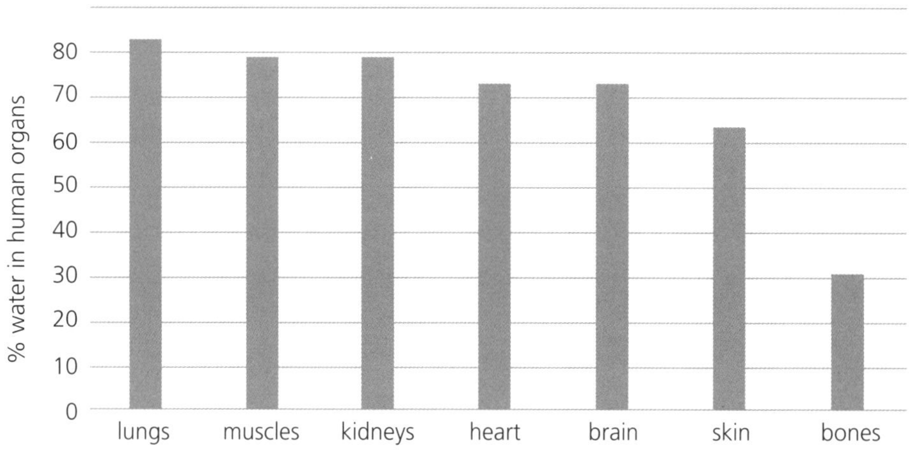

From the data in the graph, state which organ has the highest water content and which has the lowest.

....

c. Calculate the difference in water content between the organ with the highest percentage of water and the organ with the lowest percentage of water.

....

....

Extension

Each of these sentences describes one important property of water. Explain why each property is so important for living organisms.

a. Water is a good solvent – many substances dissolve in it.

b. Ice floats on top of liquid water, forming an insulating layer.

c. It takes a lot of energy to heat water up and it cools very slowly.

4.2 The water cycle

1. Label the diagram below to show the processes that take place as water changes state.

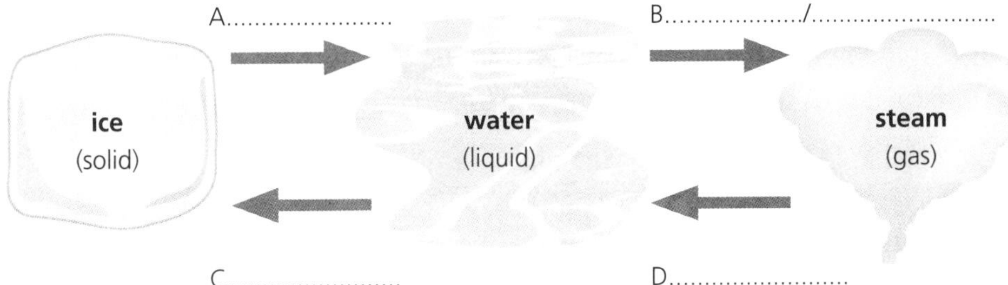

2.

This diagram shows the main stages of the water cycle. Describe what happens as water moves through the environment in the different stages of the water cycle. Answer in complete sentences.

a. Evaporation:

....

....

b. Condensation:

....

....

c. Precipitation:

....

....

d. Collection:

....

....

Extension

Rahenda tells her little brother that the water he is drinking is the same water that the dinosaurs drank. Discuss this statement and explain how it could be true.

4.3 Global warming and the water cycle

Science in context

1. Look at the following statements about global warming and the carbon cycle. Decide whether they are True or False. Put a T in the box beside each sentence you think it is true, and an F in the box beside each one you think it is false.
   - ☐ Climate change is causing problems in the water cycle.
   - ☐ Most of the water on the Earth is fresh water.
   - ☐ The uses of science always have negative environmental impacts around the world.
   - ☐ The carbon dioxide produced when fossil fuels are burned is warming the surface of the Earth and causing climate change.
   - ☐ As the surface of the Earth warms up, the polar ice-caps and glaciers around the world are melting.
   - ☐ As the surface of the Earth warms up, the climate gets drier.
   - ☐ The sea expands as it gets warmer making sea levels rise.

2. Give two ways in which the use of science in car engines and electricity generation is having a global environmental impact on sea levels.

   ....

   ....

   ....

   ....

   ....

3. a. Explain how global warming is making flooding more common in some parts of the world.

   ....

   ....

   ....

   ....

   ....

   b. Explain how global warming is making droughts more common in some parts of the world.

   ....

   ....

   ....

   ....

   ....

   ....

5.1 Diffusion in biology

1. Read the following paragraph. Complete the information using words from the box below. Each word may be used once, more than once, or not at all.

gases **net** **particles** **high** **diffusion**
lower **liquids** **down** **random** **concentration gradient**

Matter is made up of moving .... too small for us to see. A process called .... takes place in .... and .... as a result of these .... movements. Diffusion is the .... movement of .... from an area where they are at a .... concentration to an area where they are at a .... concentration. The particles move .... a ....

2. State three examples where diffusion is important in living organisms.

....

....

....

3.

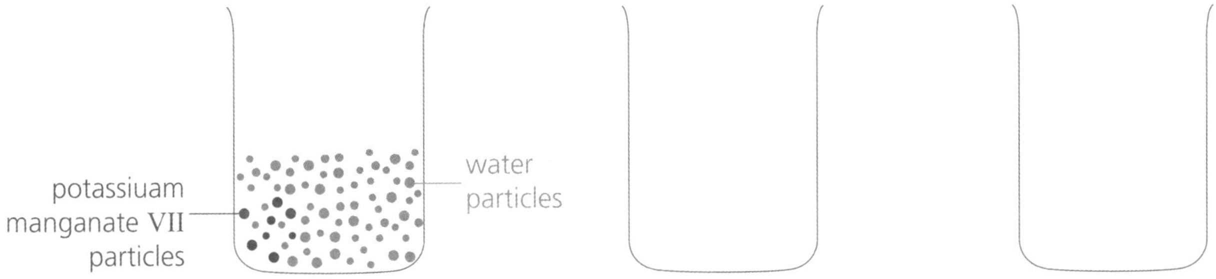

potassiuam manganate VII particles

water particles

The first diagram shows you a model of the particles when a crystal of purple potassium manganate VII is dropped into a beaker of water.

a. Draw and label a model of the particles in the beaker after 30 minutes in the first empty beaker.

b. Draw and label a model of the particles in the beaker after 24 hours in the second empty beaker.

Extension

Describe what happens in the beaker over time in terms of diffusion, the concentrations and the movements of the different types of particles.

5.2 Aerobic respiration in animals and plants

1. Read the following paragraph. Complete the information using words from the box below. Each word may be used once, more than once, or not at all.

water	**controlled**	**carbon dioxide**	**oxygen**	**energy**	**aerobic**	**glucose**

The cells of your body need .... to carry out all the processes of life. This is usually supplied by the process of .... respiration, when the .... stored in .... molecules is released in a .... way using .... The waste products of the reaction are .... and ....

2. Complete the summary equation for aerobic respiration shown below:

Glucose + ... → .... + ...

3.

A

B

a. State which of these cells is an animal cell and which is a plant cell.

....

b. Name and label the structures in both of these cells where aerobic respiration takes place

....

Extension

Explain why some cells, such as muscle cells in animals and the cells that produce seeds and fruits in plants, have many mitochondria and other types of specialised cells, such as storage cells, have very few mitochondria.

5.3 Anaerobic respiration

Extension

1. Complete the equations for aerobic and anaerobic respiration.
   a. **Aerobic:** glucose + …………… ➔ ……………. …………… + ………………
   b. **Anaerobic:** glucose ➔ ……………… ………………

2. Write T after the true statements and F after those that are false. Then write corrected versions of the statements that are false.
   a. Anaerobic respiration does not require oxygen. ………………
   b. Aerobic respiration only provides short bursts of energy. ………………
   c. Anaerobic respiration produces lactic acid. ………………
   d. Anaerobic respiration releases the same percentage of the total energy in glucose as aerobic respiration. ………………
   e. Aerobic respiration is the main type of respiration used in sprinting races. ………………

   Corrected versions of false statements:

   ……………………………………………………………………………………………

   ……………………………………………………………………………………………

   ……………………………………………………………

   ……………………………………………………………

   ……………………………………………………………

3. The diagram below shows an athlete's oxygen uptake before, during and after exercise.
   a. Add labels to show what is happening in each part of the graph.
   b. Explain why anaerobic respiration cannot be used all of the time.

   ……………………………………………………………

   ……………………………………………………………

   ……………………………………………………………

   ……………………………………………………………

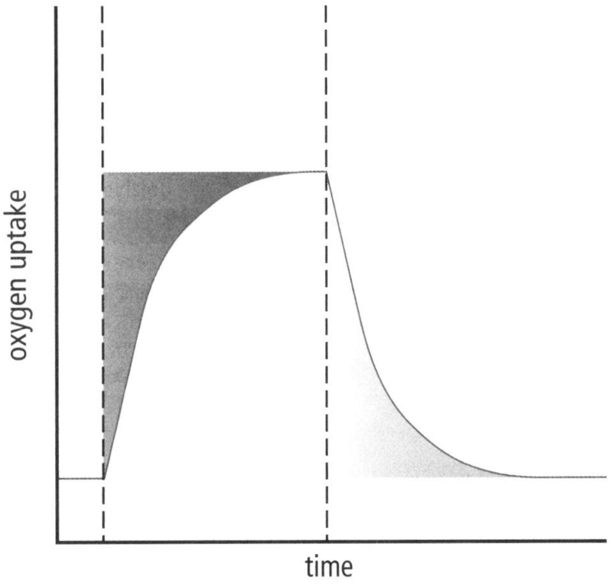

oxygen uptake

time

   c. Explain why anaerobic respiration can produce a sudden burst of energy even though it only releases a small percentage of the energy in each glucose molecule.

   ……………………………………………………………………………………………

   ……………………………………………………………………………………………

   ……………………………………………………………………………………………

   ……………………………………………………………………………………………

5.4 Investigating respiration

Thinking and working scientifically

1.

a. State what these symbols are used for.

....

b. Draw the symbol you see on a bottle of limewater and state what it indicates.

....

c. Explain how limewater is used to indicate that respiration is taking place.

....

....

2. You are asked to investigate whether a person produces more carbon dioxide after 2 minutes of exercise than when they are resting.

a. State your hypothesis.

....

....

b. Draw and label the apparatus you will use.

c. State the variables you will control and the variables you cannot control.

....

....

d. State the results you expect and explain why.

....

....

....

Extension

Scientists investigated whether germinating seeds or earthworms produce more carbon dioxide. They placed equal numbers of each in muslin bags in sealed tubes over limewater.

a. Suggest a control for these investigations.

b. Describe the results you would expect from this investigation and explain your prediction.

5.5 The lungs and gas exchange

1. Label the diagram below. Then add arrows to show how air travels to the alveoli.

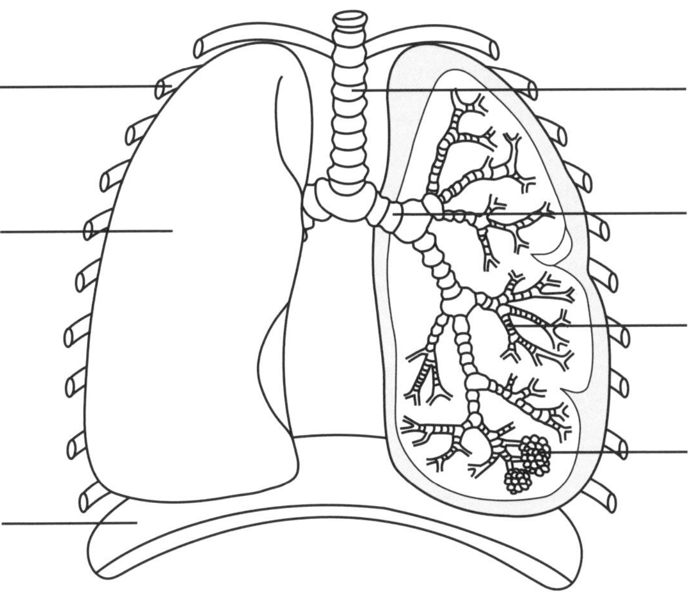

2. Read the following paragraph and fill in the gaps with words from the box below. You may use words more than once if needed.

gas exchange **carbon dioxide** **lungs** **respiratory system** **oxygen** **aerobic** **air**

The .... needed by the cells for .... respiration is brought into your body by the .... This system also removes the carbon dioxide produced during respiration. .... is moved in and out of your .... This is where .... takes place. Your body exchanges .... from the air for .... from your blood.

3. Draw lines to match the parts of the human respiratory system to their functions in the body.

Structure
A nose and mouth
B trachea
C bronchus/bronchi
D alveolus/alveoli
E pleura
F diaphragm

Function
1 tubes that carry air from the trachea into the lungs and back out again. They branch and get smaller. They are lined with ciliated cells that remove dust and harmful microorganisms from the air.
2 a sheet of muscle that divides the chest from the other organs and helps move air in and out of the lungs when you breathe.
3 slippery membranes that surround the lungs and make it easier to breathe.
4 tube with rubbery cartilage rings that carries air from the nose down to the chest and back again.
5 air moves into and out of the body here. It is warmed and made moist as it goes into the respiratory system.
6 tiny air sacs at the end of the tubes adapted for gas exchange. Oxygen moves from the air into the blood and carbon dioxide moves from the blood into the air in the alveoli.

Extension

Describe the specialised cells lining the bronchi and explain how they help to prevent lung problems.

5.6 Breathing

1. Fill in the gaps in the following paragraph with words from the box below.

exercise	oxygen	glucose	exchange	blood	cells	alveoli	energy

Gas exchange means taking oxygen into your .... and releasing carbon dioxide. It happens in your .... Respiration is the release of .... from food molecules such as .... It happens in all your .... Respiration speeds up during .... to provide your muscle cells with extra energy. Gas .... also speeds up to provide the extra .... needed for respiration.

2. Complete the table below to show the difference between the air we breathe in and the air we breathe out after gas exchange has taken place.

Gas	Air breathed in (%)	Air breathed out (%)
oxygen	20.96	
carbon dioxide	0.04	
nitrogen and other gases	79	79

3. Add labels and notes to the diagrams below to show how we breathe in and out.

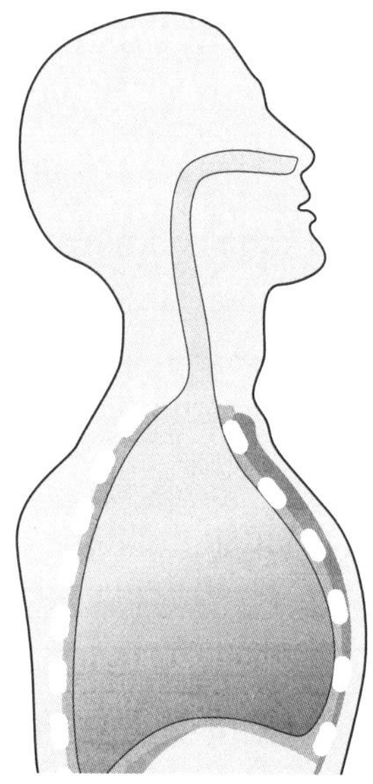

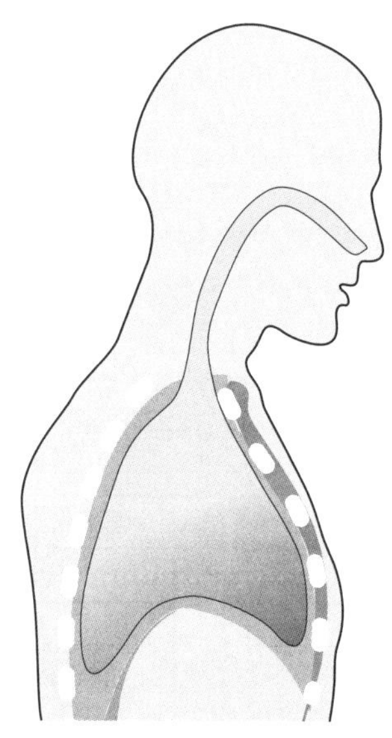

Extension

People sometimes use balloons being blown up and let down as a model of the lungs as we breathe in and out. Give two reasons why balloons are **not** a good model of the lungs and breathing.

5.7 The effect of exercise on the breathing rate

1. A teacher divides a class into Group A – students who do at least 6 hours of exercise per week and Group B – students who do less than 6 hours of exercise per week. The students investigate the effect of exercise on their breathing rate. They are going to look at two measurements:
   - the change in their breathing rate between resting and after 2 minutes of exercise
   - the time it takes for their breathing rate to go back to normal after exercise.

   a. State any difference you predict between the resting breathing rates of Groups A and B.

   ....

   ....

   b. Describe how you would expect the breathing rates of the groups to differ after exercise.

   ....

   ....

   c. Use your scientific knowledge to explain your prediction

   ....

   ....

2. The mean time for the breathing rate of Group A to return to normal after 2 minutes exercise was 2.0 minutes. For Group B, it was 3.5 minutes.

   a. Draw a bar chart to show these results.

   b. Describe the pattern you see in these results.

   ....

   c. i. State one conclusion you draw from these results.

   ....

   ii. Discuss two limitations in these results.

   ....

   ....

Extension

One student in Group A took 3.0 minutes for their breathing rate to recover. In Group B, one student recovered in 1.8 minutes, and another took 4.5 minutes. Explain how these results differ so much from the group means.

5.8 The structure of the alveoli

1. Fill in the gaps in the following paragraph with words from the box below. You may use each word once, more than once, or not at all.

   air   alveoli   capillaries   blood   oxygen   adaptations   gas exchange   diffuses   dioxide

   The lungs are the site of .... which takes place in the .... There are millions of these tiny sacs which have a number of .... so as much oxygen .... from the .... to the .... as possible. Carbon .... diffuses from the blood in the .... that surround the .... to the air inside them.

2. These diagrams show a group of alveoli in the lungs and a cross section through a single alveolus.

   Using appropriate colours if possible, on both diagrams:

   a. Add arrows and labels to show deoxygenated blood arriving from the heart.
   b. Add arrows and labels to show oxygenated blood leaving to travel back to the heart.
   c. Add arrows and labels to show air moving into and out of the alveoli.
   d. On diagram B, add arrows to show the direction of diffusion of oxygen and carbon dioxide between the blood and the air.

3. Two of the main adaptations of the alveoli for gas exchange are listed here. Explain how each adaptation makes the diffusion of oxygen and carbon dioxide more efficient.

   a. Many tiny air sacs:

   ....

   ....

   b. The walls of the capillaries and the alveoli are very thin:

   ....

   ....

Extension

Explain how the following adaptations seen in the alveoli of the lungs make gas exchange more efficient:

a. steep concentration gradients between the air in the alveoli and the blood in the capillaries
b. a rich blood supply.

5.9 Asthma

Science in context

1. a. Describe what happens in the respiratory system when someone has an asthma attack.

....

....

....

b. Explain how these changes make the person affected feel breathless.

....

....

....

....

2. Two students breathe out as fast as they can. A machine measures the volume of air they breathe out. The results are shown on the graph.

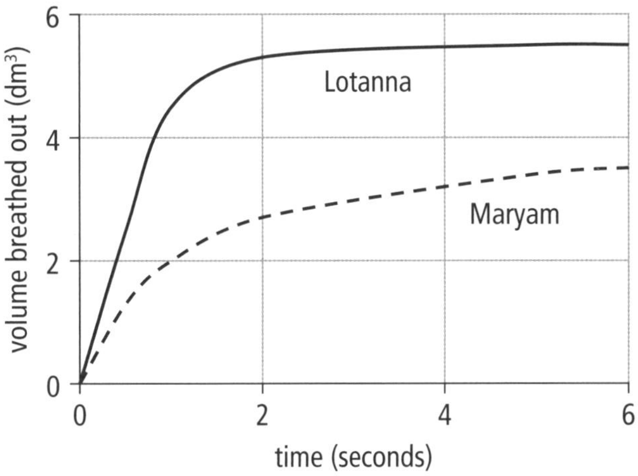

a. Describe two differences between Lotanna and Maryam's results

....

....

....

....

volume breathed out (dm^3)
6
4
2
0
0
2
4
6
time (seconds)
Lotanna
Maryam

b. Deduce which student has asthma.

....

3. Most people with asthma now live healthy lives and are able to take part in sporting activities as a result of several different medicines.

Explain how scientific knowledge had to be applied to develop and deliver asthma medicines so they work well.

....

....

....

....

....

5.10 Transport in the blood

1. **a.** State the four main components of blood.

   ....

   b. State which type of blood cell is most common and explain why.

   ....

   ....

2. Statements **a–f** refer to the four main components of blood. Write the name of the correct component next to each statement.

   a. Carries oxygen around the body ....

   b. Needed to fight infections ....

   c. Packed full of haemoglobin ....

   d. The largest cells in your blood ....

   e. A pale yellow liquid ....

   f. Carries dissolved substances around your body ....

   g. Helps to form a clot when a blood vessel is damaged ....

   h. Has a biconcave shape to increase its surface area ....

3. These data show the proportions of the different components of your blood.

Component of the blood	Plasma	Red blood cells	White blood cells and platelets
% volume of the blood	55	44	1

Show these data as a simple bar chart.

Extension

Anaemia is a condition when a patient does not have enough red blood cells. Three of the symptoms of anaemia are listed here. Use your knowledge of red blood cells to explain these symptoms.

a. Anaemia makes you feel tired all the time.

b. A patient with anaemia has less haemoglobin than normal.

c. Anaemia can be caused by a lack of iron in the diet.

5.11 Coping with extremes

Extension

1. At high elevations, air pressure is lower making it harder to get oxygen into your body. People born at high elevations have several adaptations which help them survive in this difficult environment.
   a. List four adaptations commonly seen in the respiratory systems of people born at high elevation.

      ....

      ....

      ....

   b. Explain how each of these adaptations makes it easier to survive at high elevation.

      ....

      ....

      ....

      ....

2. This graph shows the heart rate of a young seal during a dive.

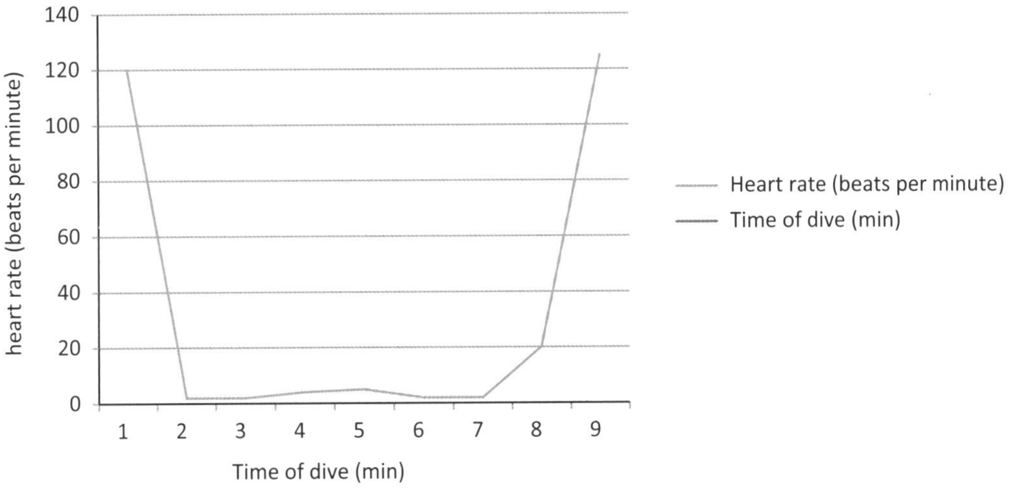

   a. Describe the adaptation for diving shown in this data.

      ....

   b. Explain how this adaptation helps the seal dive for food successfully.

      ....

      ....

   c. Describe two more common adaptations seen in diving mammals. For each, explain how the adaptation helps the animals survive.

      ....

      ....

      ....

6.1 The food we eat

1. Draw lines to match each nutrient to its role in your body:

Nutrient
A carbohydrates
B fats and oils
C proteins
D vitamins
E minerals

Role
1 help cells function properly and strengthen bones and teeth
2 help chemical reactions take place in your cells
3 most of our energy intake should come from this group
4 used to build cell membranes and a good source of energy
5 essential for growth and for repairing cells

2. Write the names of the main nutrients each of these foods contain. Choose up to two nutrients from carbohydrates, proteins and fats.

   a. Sugar ....

   b. Olive oil ....

   c. Chocolate ....

   d. Nuts ....

3. Rashid thinks that fat is bad for you. Give three reasons why we need fat in our diet.

   ....

   ....

4. Complete the following statements about digestion.

   Whatever food you eat will be broken down in your ....

   Most of your food is made up of ...., ... molecules which your body cannot use.

   In your digestive system, these molecules are broken down into ...., ... molecules which move into your blood by ....

   The dissolved nutrients are carried in your blood to the .... where they are needed.

Extension

In many parts of the world, children eat mainly rice and vegetables. Name one major nutrient they will lack and explain what problems this might cause.

6.2 Carbohydrates, fats, and energy

1. Proteins, lipids and carbohydrates are large molecules made by joining smaller molecules together. Complete this table to show the different types of small molecules that make up the main nutrients we need.

Nutrient	Small molecules joined to make them
starches	
proteins	
fats	

2. Some of these statements are true and some are false. Write T after the true statements and F after those that are false. Then write corrected versions of the false statements.

   a. Glucose is an example of a starch. ....

   b. Glucose is a small, soluble carbohydrate molecule. ....

   c. Fats and oils contain less energy per gram than carbohydrates. ....

   d. Each fat or oil has a different set of amino acids attached to a glycerol molecule. ....

   e. Starch is made from lots of sugar molecules joined together. ....

   Corrected versions of false statements:

   ....

   ....

   ....

3. Look at the following statements and use them to complete the table comparing the human storage molecules glycogen and fat.

   a short-term carbohydrate energy store

   broken down for energy over time if not enough food is eaten

   stored in special cells under the skin and around the body organs

   a long-term energy store

   used for quick, instant energy

   stored in our muscles and liver

Glycogen	Fat

Extension

Foods such as chips, burgers, ice cream, pastries, and cakes are linked to a global epidemic of people who are overweight. Discuss why foods like these are often linked to weight gain in people.

6.3 Measuring the energy in food – managing variables

Thinking and working scientifically

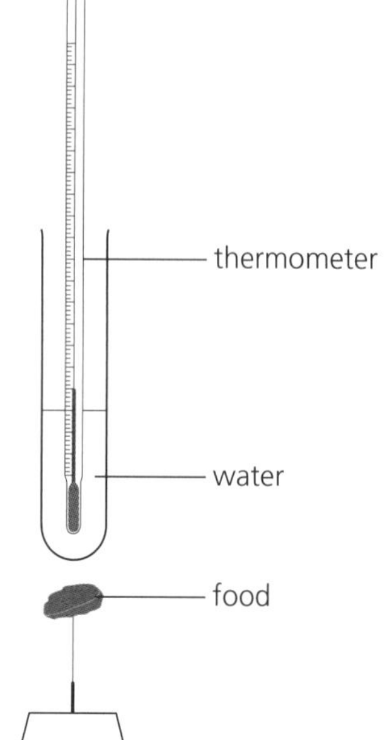

1. Eniola is testing foods to see which one contains most energy. She uses the apparatus shown on the right.

   During the investigation, variables need to be 'changed', 'measured', 'controlled' or 'calculated'. Write the correct choice next to variables **a**–**g**.

   a. Type of food ....
   b. Mass of food ....
   c. Volume of water ....
   d. Distance between food and test tube ....
   e. Water temperature before food burns ....
   f. Water temperature after food burns ....
   g. Temperature rise ....

2. Eniola uses three measuring instruments during the investigation. Complete the table to show what each instrument measures and the units.

Measuring instrument	Variable measured	Units used
thermometer		
measuring cylinder		
electronic balance		

3. When Eniola burned twice as much bread, the temperature rise of the water doubled. Give two reasons why Eniola should only burn small quantities of food.

....

....

....

....

Extension

Eniola compared the energy content of bread, chicken breast, and cheese.

Food tested	Mass of food (g)	Temperature before (°C)	Temperature after (°C)	Temperature rise (°C)	Temperature rise per gram (°C per g)
bread	1.5	21	51		
chicken	1.0	22	68		
cheese	2.0	22	88		

a. Complete the table of results and suggest a reason for the differences between the foods.
b. Suggest two different variables that Eniola could control to make her results more reliable in future.

6.4 A balanced diet

1. Read the following paragraph and fill in the gaps with words from the box below.

lipids	**nutrient**	**fibre**	**energy**	**water**	**minerals**	**rice**	**proportions**	**proteins**

Each word may be used once, more than once, or not at all.

A balanced diet contains every essential ...., in the correct .... These include ...., ..., vitamins, and .... You also need starchy carbohydrates like .... to supply .... and .... and .... to keep your body and digestive system working well.

2. The table shows the nutrients in bars of milk chocolate.

Nutrient	Mass present (g/100g)
sugar	57
protein	8
fat (mostly saturated)	30

Give two reasons why chocolate should not be your main source of carbohydrate.

....

....

....

3. a. Describe what is meant by the food pyramid

....

b. In the food pyramid below, fill in the type of foods that should be in each level of the pyramid and give one example of each.

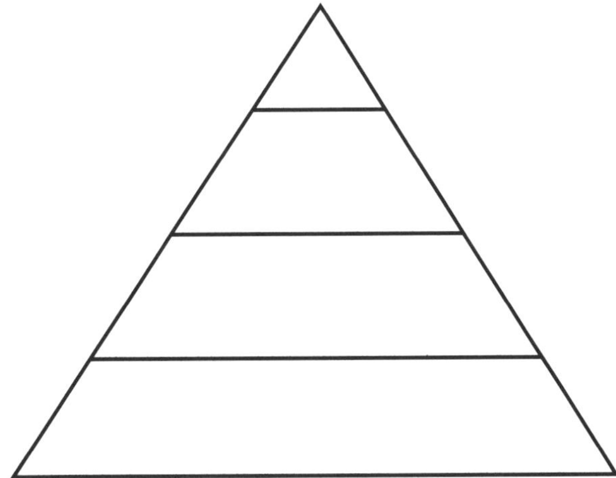

Extension

The balanced diet you need changes through your life. State three different factors that would change the balanced diet you need during your life, and explain why those factors affect the food you need to eat.

6.5 Diet, growth, and development

1. Read the following paragraph and fill in the gaps with words from the box below. Each word may be used once, more than once, or not at all.

scurvy	nutrients	explain	vitamins	lack	ill	deficiency

In the past, poor diets made many people .... Scientists used creative thinking to .... why this happened. They suggested that illnesses like .... were .... diseases. They were caused by a .... of essential .... or minerals. When the missing .... were supplied, people recovered.

2. Draw lines to match each of these deficiency diseases to the missing nutrient.

Deficiency disease
anaemia
night-blindness
scurvy
rickets

Missing nutrient
vitamin C
calcium
iron
vitamin A
vitamin D

3. Hassina does not feel well. Her doctor tests her blood. The results are shown below:

Vitamin/mineral	Normal blood levels (units)	Hassina's blood (units)
calcium	2.2–2.6	2.4
vitamin C	0.4–1.5	0.2
iron	50.0–170.0	30.0
vitamin A	30.0–65.0	55.5

a. List any deficiency diseases Hasina may have.

....

....

b. State the symptoms you would expect her to have and suggest two different foods she should eat in future to make sure she does not become ill again.

....

....

....

....

Extension

a. Plot separate bar charts to show the normal maximum and normal minimum levels of the vitamins and minerals shown in this table. Add Hassina's blood readings to each graph.

b. Explain why it is important to plot each different vitamin and mineral on a different bar chart.

c. Explain how displaying data on a bar chart like this can make trends and patterns in Hassina's blood results easier to identify.

6.6 Starvation, obesity, and health

1. Underline the correct answers from these pairs.

If you do not get enough food, you become **too thin / too fat**. Your muscles **bulge / waste away**, you have **little / lots of** energy, and you get **excess / deficiency** diseases. Starving children **thrive / fail to grow** and they **do / do not** develop healthily. **Starving / Overfed** adults cannot work and become **healthy / ill.** Starvation **is / is not** a lifestyle choice – about 9 **million / thousand** people die of starvation every year.

2. Read the following paragraph and fill in the gaps with words from the box below.

heart	blood	fat	diabetes	body	energy	obese

In most countries, average .... masses are rising and more people are becoming .... Their diets supply more .... than they use and the excess is stored as .... This can cause long-term health problems like ...., cancer, high .... pressure, and .... disease.

3. Many health workers want to help people improve their diets. Match each action they propose with the correct reason.

Action
A Make fatty and sugary foods more expensive
B Add extra nutrients to common foods
C Encourage people to exercise more
D Add extra genes to common crops

Reason
1 so people automatically get a balanced diet when they eat everyday foods.
2 so they use more energy and are less likely to become obese.
3 so people buy fewer of the foods high in fats and sugars which easily lead to obesity.
4 so people automatically get a balanced diet when they eat everyday foods.

4. Obesity is linked to an increased risk of getting many diseases. Describe the main symptoms of the three obesity-linked diseases listed here:

Diabetes: ....

....

Heart disease: ....

....

Arthritis: ....

....

6.7 Smoking and health

1. The diagram below shows the cells that line your bronchi, the tubes that lead to your lungs.

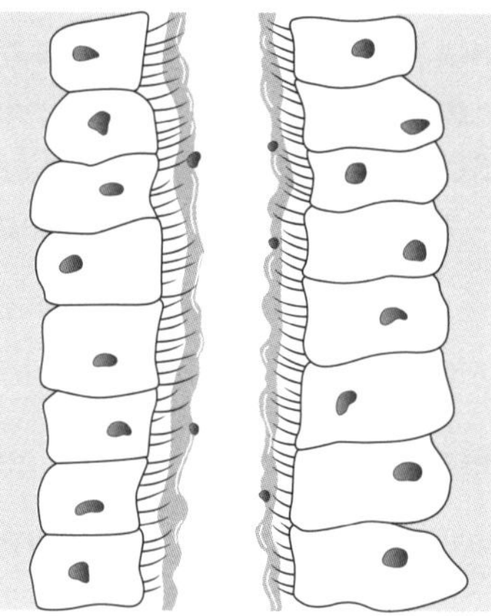

a. Add labels to describe the adaptations in these specialised cells that help to keep the alveoli of your lungs clean.

b. List six changes that take place in your body if you smoke tobacco.

....

....

....

....

....

....

c. Name two diseases that smokers are more likely to suffer from than non-smokers.

....

....

2. Fill in the gaps in the following paragraph with words from the box below.

raises **cancer** **addictive** **narrower** **harmful** **reduces** **heart** **cilia**

Cigarette smoke contains many .... chemicals. Tar paralyses .... and makes mucus build up in the lungs. It is also carcinogenic so it can cause lung .... Carbon monoxide .... the amount of oxygen blood can carry. Nicotine .... smokers' blood pressure by causing blood vessels to become ...., so it increases their risk of having a .... attack. It also makes cigarettes ....

Extension

Two friends run a 200 m race. Neither of them does much exercise and one smokes 20 cigarettes per day. Their results are shown in the table.

Runner	Race time (seconds)
Imran	29
Mohamed	45

Identify which of the friends is a smoker and explain why his race time is so different.

6.8 Building the evidence

Thinking and working scientifically

1. Fill in the gaps in the following paragraph with words from the box below. You may use each word once, more than once, or not at all.

link	risk	tobacco	peer-reviewed	lifestyle	observed
smoking	results	lung cancer	smokers	evidence	causal link

People have smoked and chewed .... for centuries. The link between .... and .... was first .... in the mid-20th century by Sir Richard Doll. He collected .... information from his lung cancer patients and noticed that most of them were .... He published his .... in a .... journal. Scientists in many other countries collected more .... which confirmed the .... between smoking .... and an increased .... of developing lung cancer.

2.

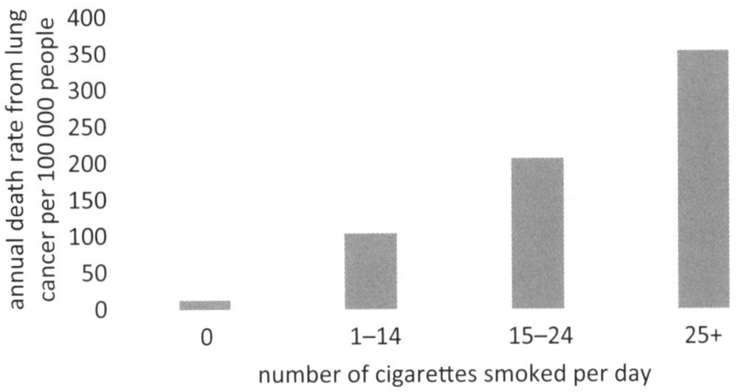

a. Using the data on this graph, calculate the increased risk of developing lung cancer in someone who smokes more than 25 cigarettes a day compared to someone who does not smoke. Show your workings.

....

....

....

b. Discuss what the data in the graph in question 2 tells you about the link between cigarette smoking and the risk of developing lung cancer.

....

....

c. Doll published his original findings in a *peer-reviewed* journal. Explain why this is important. ....

....

....

....

d. State two things you need to know about the quality of the evidence from the graph in question 2, and explain why they are important.

....

....

6.9 The human skeleton

1. This diagram is of an adult human skeleton. You are also given a list of the names of some of the bones which make up the skeleton. 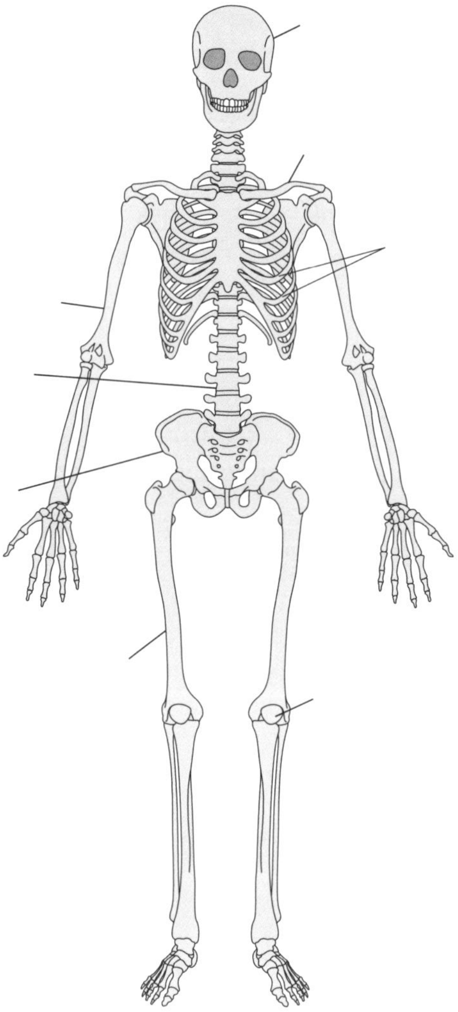
   - **a.** skull
   - **b.** ribs
   - **c.** vertebral column
   - **d.** pelvis
   - **e.** collar bone
   - **f.** femur
   - **g.** humerus
   - **h.** kneecap

   a. State four functions of the skeleton in your body.

   ....

   ....

   ....

   ....

   b. Choosing from the list of bones given, label one bone or area of the skeleton which carries out each of the functions you state in part **a**.

Extension

Draw a diagram to show the structure of bone. Annotate your diagram to explain how the different parts of the bone are adapted to carry out their function.

6.10 Muscles and movement

1. The diagram shows a human leg and pelvis and the joints at each end of the femur (A and B).

a hinge joint	**a ball-and-socket joint**	**connects leg to hip**
lets leg swing freely	**lets leg bend and straighten**	**found in the knee**

Match each of the facts below to the correct joint – write them in the correct column of the table.

Joint A	Joint B

2. Each joint contains different tissues. Link each tissue to its function in the joint with a line.

Tissue
A ligament
B cartilage
C bone
D synovial fluid

Function in the joint
1 prevents the ends of the bones from rubbing together in the joint
2 provides support and moved by muscles
3 holds the bones together but allows them to move within the joint
4 lubricates joints so bones slide over each other smoothly

3. Four students write about antagonistic muscles.

Leah Most movements are caused by antagonistic muscles.

Acho There are pairs of antagonistic muscles in your arms and legs.

Karis When one muscle contracts to pull a bone, the antagonistic muscle relaxes.

Mikayla Muscles work in pairs because they can only pull bones in one direction.

a. Give the name of the student who explained **how** antagonistic muscles work. ....

b. Name the student who explained **why** antagonistic muscles work like this. ....

4. Diagram A shows the pair of antagonistic muscles that bend your arm. Add the muscles that straighten your arm to diagram B. Label the muscles with their name and state whether they are contracted or relaxed.

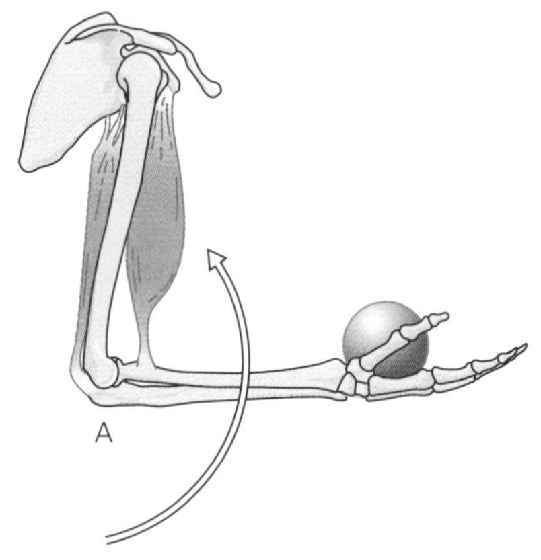

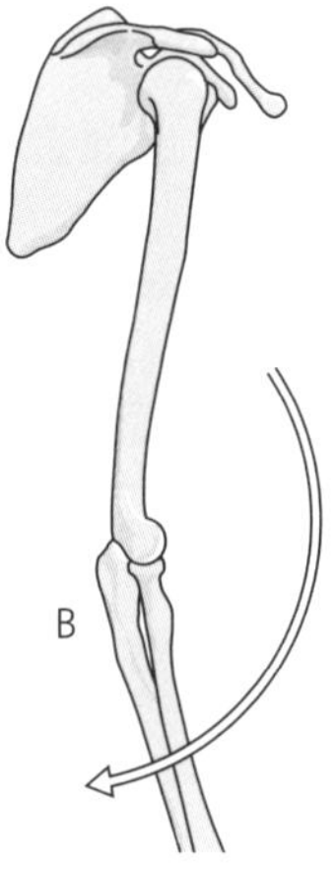

Science in context

1. Read this extract from a magazine article. Write two more paragraphs, explaining some of the ways in which we can prevent diseases linked to food. Make sure you discuss the importance of scientific understanding in tackling these problems.

You are what you eat

Have you had a good meal today? Sadly, for 11% of the world population, the answer to that question is 'no'. Millions of people suffer deficiency diseases because they do not get enough to eat.

Millions more suffer from diseases which result from being overweight – obesity causes deaths from diabetes, heart disease and some cancers.

We have more scientific knowledge about diet, health, and plant growth than ever before. We have fertilizers and specially bred crops which yield plenty of food. We have better ways of controlling plant diseases and the pests which destroy crops than at any time in human history. We have ways of transporting food around the world.

How can we prevent people dying because they eat too much or too little food?

7.1 Ecosystems of the Earth

1. Read the following paragraph and fill in the gaps with words from the box below.

Each word may be used once, more than once, or not at all.

biosphere	**biotic**	**ecosystem**	**temperature**	**light**	**environment**	**abiotic**	**soil**

An .... is an .... that supports life. The biggest .... is the Earth itself, known as the .... Each .... is made up of non-living or .... factors and of living or .... factors. Non-living factors include the ...., the ...., the amount of ...., and the water supply.

2. State three biotic factors that affect the organisms in an ecosystem.

....

....

....

3. Match each of the ecosystems below to the correct description:

Ecosystem
A desert
B coral reefs
C ocean
D temperate woodland
E tropical rain forest

Characteristics
1 large changes in light and temperature levels through the year
2 a lot of rain, a lot of sun, and warm temperatures all year round
3 dry areas with little rainfall and little available water – may be hot or cold
4 salty water with a relatively stable temperature, always moving
5 warm, clean, relatively shallow seas and high biodiversity

Extension

a. Define biodiversity.

b. Describe the level of biodiversity you would expect to see in each of the ecosystems given in question 3.

7.2 Habitats within an ecosystem

1. Write ‘desert’ or ‘rainforest’ under each animal to show the habitat it is adapted for.

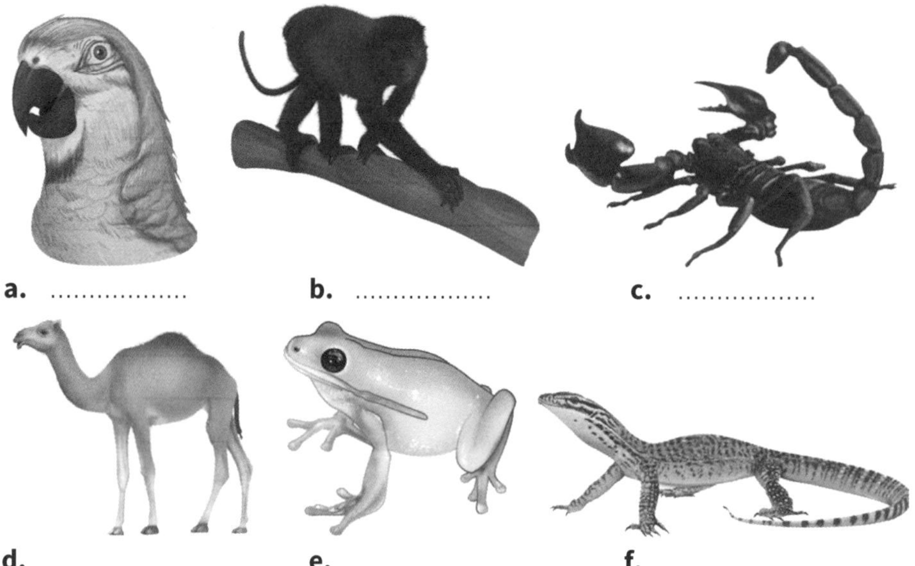

a. ....   b. ....   c. ....

d. ....   e. ....   f. ....

2. State the habitat where adaptations **a** to **g** would be useful: in a desert, a rainforest, or the Antarctic. Some may be useful in more than one place.
   a. A thick layer of fat under the skin ....
   b. Ability to live in underground tunnels ....
   c. Ability to climb easily ....
   d. Large energy stores ....
   e. Large ears ....
   f. Strong arms ....
   g. Wide feet ....

3. The Arctic is a cold and difficult habitat. There is a lot of ice and snow. The temperatures are very low. Polar bears are predators who feed on seals and other animals in the Arctic. Look at this drawing of a polar bear. State three adaptations **visible in this drawing** and explain how they help the polar bear to survive in its Arctic habitat.

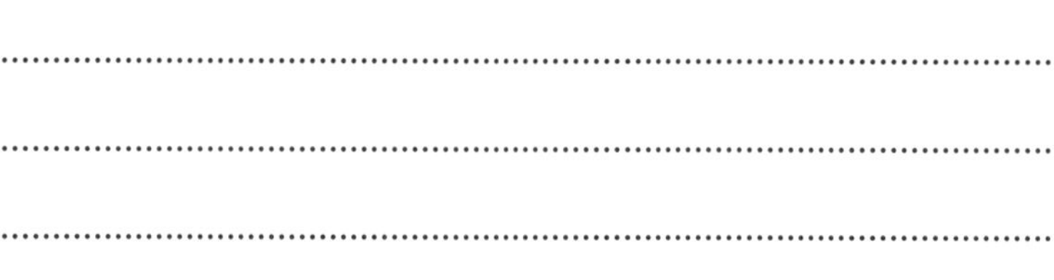

Extension

a. Describe the main characteristics of your own local habitat.
b. Describe three adaptations seen in local animals or plants and explain how they enable them to survive in the habitat.

7.3 The interdependence of organisms

1. Draw lines to match each word to the correct definition:

Word
A predator
B prey
C consumer
D producer
E carnivore
F scavenger
G herbivore

Definition
1 eats other living things to obtain nutrients
2 only eats animals that are already dead
3 only eats animals
4 only eats plants
5 makes its own food
6 is hunted and eaten by other animals
7 hunts and eats other animals

2. State the meaning of the following terms:

Population ....

Interdependent ....

3. Food chains like the one below are found on many grasslands.

producer → herbivore → carnivore

a. Give an example of a food chain like this.

....

b. Describe what you expect to happen to the herbivore and producer populations if the carnivore numbers are reduced by hunting and explain why.

....

....

....

c. Suggest two reasons why the herbivore population might fall in the future.

....

....

Extension

In parts of the Arctic circle, the populations of wolves and caribou are interdependent. Add these labels to the graph to explain the changes in wolf and caribou populations shown.

caribou eaten by wolves; wolf numbers drop because there are fewer caribou to eat; caribou raise offspring; wolves kill more caribou, so they breed successfully and numbers rise; the cycle repeats.

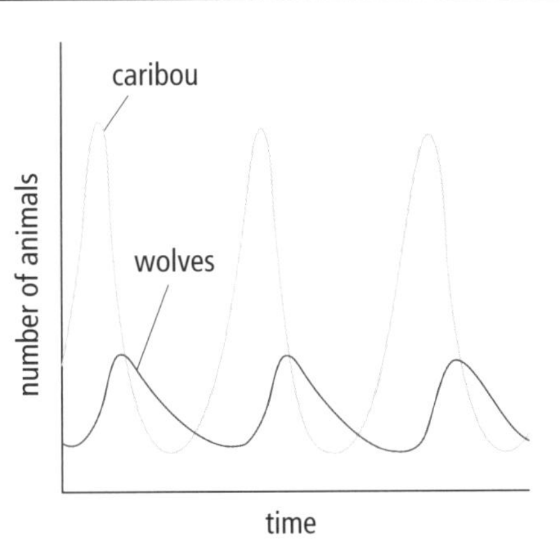

7.4 Pesticides and bioaccumulation

1. Scientists investigated bioaccumulation in a fish-eating bird's food chain.

Table 1 shows their results. Table 2 shows the results of another investigation into the same birds.

Table 1

Animal	Concentration of pollutant (ppm)
fish-eating bird	25.0
large fish	2.0
small fish	0.5
invertebrates	0.1
water plants	0.001
water	0.00001

Table 2

Concentration of pollutant (ppm)	Eggshell thickness (mm)
10	0.24
7	0.25
6	0.26
5	0.26
4	0.27

Use this information to:

a. Draw the food chain being affected by bioaccumulation.

....

b. Explain why the birds of prey contain so much of the pollutant.

....

....

....

c. In areas with high levels of water pollution, fish-eating birds are found dead. In areas with slightly lower levels of pollution, fish-eating birds don't die but the eggs often break when the chicks are developing so the birds fail to reproduce. Suggest reasons for both of these observations.

....

....

....

....

Extension

a. Display the results from Table 2 as a line graph and describe what it shows.

b. Explain why the data from Table 1 would be difficult to display as a normal bar chart or line graph.

7.5 Invasive species

1. Draw lines to match each of these terms to their definition:

Ecological term
A native species
B introduced species
C invasive species
D biological pest control

Definition
1 species introduced into an ecosystem by humans which then damages the ecosystem
2 organisms introduced to an ecosystem to control invasive species
3 species found naturally in a balanced ecosystem
4 species brought into an ecosystem by humans that do not cause any harm

2. Read the following paragraph and ring or underline the correct word from each pair.

A **native** / **invasive** species is a type of animal or plant introduced into an **ecosystem / garden** by humans, which then **improves / damages** the ecosystem. **Introduced / invasive** species often **reproduce rapidly / fail to reproduce.** A **new / old** species has **many / no** predators so the **individual / population** grows fast. The invasive species often **breaks / supports** existing food chains and webs within **a rainforest / an ecosystem.**

3. Lionfish are an invasive species – they are relatively big, carnivorous fish that have been introduced in many areas of the Atlantic ocean. In several shallow areas, they have wiped out up to 95% of the native fish. They have no natural predators. Scientists discovered lionfish taste good, and they have shown people how to catch and cook lionfish. Now people are being used as biological pest control to keep lionfish populations small. Scientists have discovered that lionfish stop causing problems to native fish if lionfish populations are kept small.

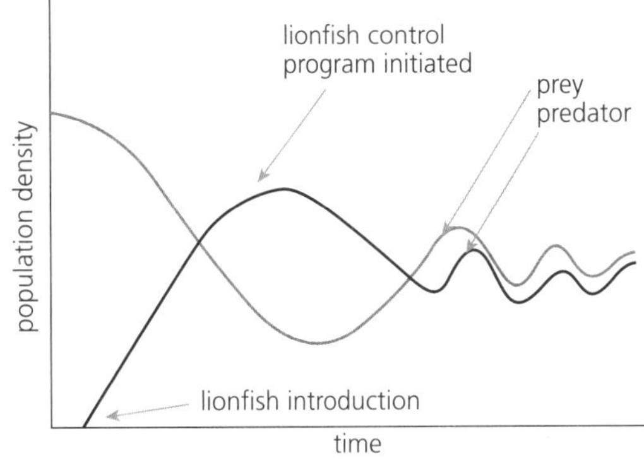

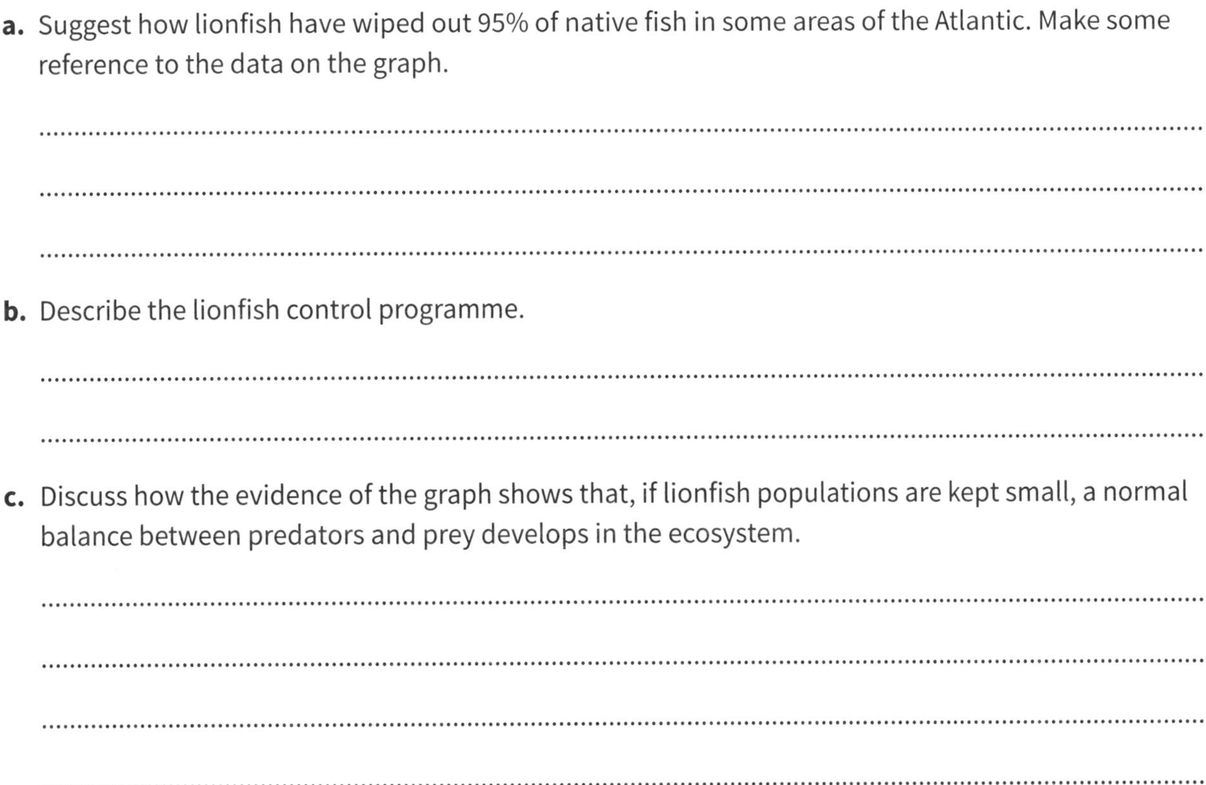

a. Suggest how lionfish have wiped out 95% of native fish in some areas of the Atlantic. Make some reference to the data on the graph.

....

....

....

b. Describe the lionfish control programme.

....

....

c. Discuss how the evidence of the graph shows that, if lionfish populations are kept small, a normal balance between predators and prey develops in the ecosystem.

....

....

....

....

7.6 Invasive species and ecosystems

1. Fill the gaps in the following paragraph with words from the box below.

predators	quickly	extinct	food	invasive	disease

Population sizes are normally controlled by the amount of .... available, the number of predators, and .... If new .... appear, or the food supply drops, a species can become .... Many extinctions are caused by .... species. These grow and reproduce .... and have few predators.

2. Brown tree snakes have caused big changes on the island of Guam.

   a. Link these species to show the food webs present before and after the arrival of the tree snakes.

before: lizards, **insect-eating birds**, spiders, **seed- and fruit-eating birds**, insects, bats, leaves, fruit, and seeds from a wide range of plants

after: brown tree snakes, lizards, spiders, insects, bats, leaves, fruit, and seeds from a wide range of plants

   b. State whether the numbers of the species listed below went up or down after brown tree snakes arrived on Guam, and explain each answer.

Insect-eating birds..

Lizards ....

....

Spiders..

Bats..

Seed and fruit-eating birds..

   c. How will the forest be affected if the insect population increases?

....

....

Extension

Insect-eating birds cut the number of insects feeding on trees. Seed- and fruit-eating birds spread tree seeds and make them more likely to germinate and grow. Both types of birds have become extinct on the island of Guam as a result of the brown tree snakes. Predict, with reasons, the effect that the loss of both types of birds will have on the forest.

7.7 Sampling your ecosystems

Thinking and working scientifically

1. Read the following paragraph and fill in the gaps with words from the text box. Each word may be used once, more than once, or not at all.

populations **calculated** **quadrats** **sampled** **species** **sampling** **estimate**

As the human population has risen, the numbers of other .... have fallen. Scientists .... how many plants or animals of a particular species there are by .... their .... Plants and slow-moving animals are sampled using .... and other types of animals are .... using traps. Using these methods, an estimate of the whole population is ....

2. Jaz is a farmer who looks after his field, but he has found thistles are starting to grow in the grass. Thistles are prickly plants which animals like sheep, cows, and horses do not like eating. Jaz thinks these thistles are spreading from his neighbour's field. Tun, his neighbour, claims he does not have a thistle problem and that all fields have lots of thistles. Tun owns the field where quadrats A and C are taken. Jaz owns the field where quadrats B and D are taken.

A B C D

Some students carry out sampling in both fields to investigate these claims – see quadrats A–D.

a. Describe a quadrat. ....

....

b. Describe how a quadrat is used. ....

....

c. Using the 0.25 m^2 quadrats shown, calculate the mean number of thistles per m^2 in each farmer's fields. Show your working, including how you count the number of thistles in each quadrat.

....

....

....

....

....

....

d. Based on your results, evaluate the claims of the two farmers and decide who you think is correct.

....

....

e. Suggest one way to improve the investigation. ....

....

8.1 What is weather?

1. Read the following paragraph about the weather and fill in the gaps with words from the box below. Each word may be used once, more than once, or not at all.

air pressure	**hail**	**humidity**	**snow**	**weather**	**wind**
precipitation	**sleet**	**atmosphere**	**temperature**	**clouds**	

The .... we get on the ground is the result of factors in the .... Changes in .... affect .... strength and direction. The .... (amount of water in the air) and the air .... affect the formation of .... and the amount of .... we get. There are four main forms of .... – rain, ..., ..., and ....

2. **a.** Define the term 'weather'.

....

....

3. Complete this table which shows some commonly used weather symbols.

Symbol	Weather
	sun

	snow

Extension

Explain why weather varies so much across a country, or in small areas of a country.

8.2 Climate and climate change

1. **a.** Define the term 'climate'.

   ....

   ....

   b. Describe the similarities between weather and climate.

   ....

   ....

   c. Explain the difference between weather and climate.

   ....

   ....

   ....

   ....

2. Match each climatic zone to the conditions you expect to see there.

Climatic zone
A arid
B continental
C mediterranean
D polar
E temperate
F tropical

Conditions
1 mild summers and cool winters
2 hot and humid with a lot of rainfall and big storms
3 extremely long, very cold winters and short, cold summers
4 very dry, with very little rain
5 short, hot summers and long, cold winters
6 hot, dry summers and cool, wet winters

3. The climate of the earth has changed many times over millions of years.

   a. Give one piece of evidence that shows the climate of the Earth has changed over millions of years.

   ....

   ....

   b. Give one piece of evidence that the climate of the Earth is changing today.

   ....

   ....

8.3 Climate changes past and present

1. Here are four possible causes of global climate change. Explain how each of these would affect the temperature at the surface of the Earth.
   a. The orbit of the Earth around the Sun: ....

   ....

   b. Volcanoes erupting and big meteorites hitting the surface of the Earth: ....

   ....

   c. Changes in the activity of the Sun: ....

   ....

   d. Changes in the levels of carbon dioxide in the atmosphere: ....

   ....

2. This graph shows natural cycles of climate change over thousands of years.

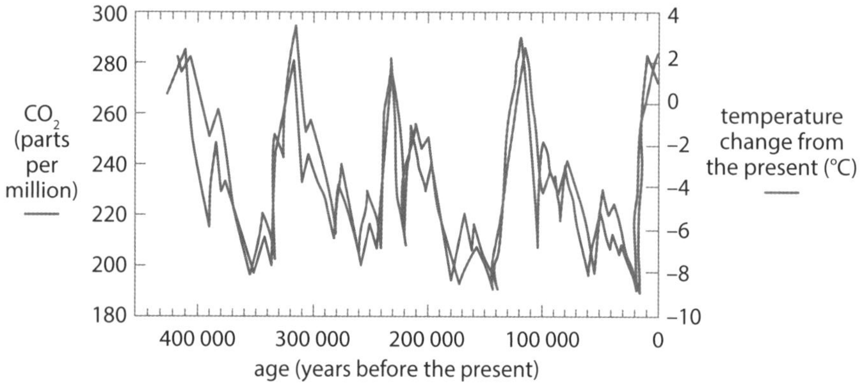

   a. Describe natural climate change. ....

   ....

   b. Explain what an ice age is and label an ice age on the graph.

   ....

   c. State the concentration of carbon dioxide in the atmosphere in ppm during this ice age.

   ....

   d. Identify and label a period of high global temperatures on this graph.

   e. State the concentration of carbon dioxide in the atmosphere in ppm during this period of high global temperatures. ....

   f. Discuss why the data on this graph is useful to scientists today.

   ....

   ....

Extension

Explain how climate change today is similar to and different from the patterns seen in the graph for question 3.

8.4 Gathering evidence of climate change

Science in context

1. Scientists use evidence from the past to compare the amount of carbon dioxide in the atmosphere with Earth's temperature. Complete the table to show what this evidence shows.

Measurements taken	What the evidence shows
past temperature readings from weather stations all over the world over many years to the present	
the thickness of tree rings on very old trees, showing how fast the trees grew each year	
the concentration of carbon dioxide in atmospheric air trapped in layers of ice thousands of years ago	

2. Scientists analyse air trapped in ice cores from thousands of years ago to get data on changes in the carbon dioxide concentrations over time.

Thousands of years ago	Carbon dioxide (ppm)
100	220
80	250
60	200
40	190
20	180
0	420

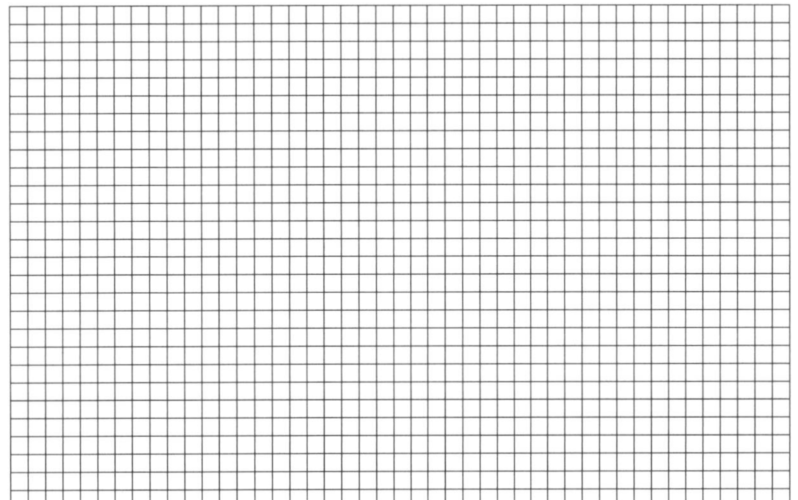

a. Use these data to draw a line graph (make the *x*-axis run backwards from 100–0 thousand years ago).

b. Compare carbon dioxide levels now (time 0) and 20 000 years ago. State how much extra carbon dioxide is in the atmosphere now. ....

Extension

Suggest why scientists are so worried about the pattern shown in the graph you drew to answer question 2a.

9.1 What do we know about plants?

1. **a.** The diagram shows a simple flowering plant. Label its four main parts.

   b. Give the main functions of the plant parts labelled A–D.

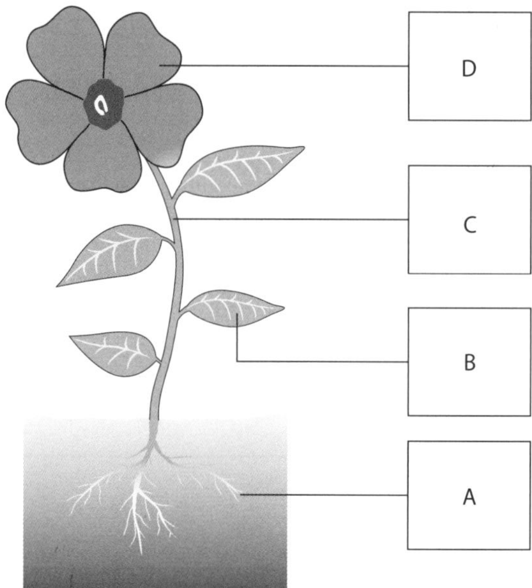

A ....

....

B ....

....

C ....

....

D ....

....

2. List three reasons why plants are such important organisms and explain each one.

....

....

....

....

....

....

Extension

The stems of cacti are swollen and green. The leaves are sharp spines. State the functions of these specialised plant parts and suggest how they are adapted to carry out their functions.

9.2 Photosynthesis

1. **a.** Describe the process of photosynthesis.

   ....

   ....

   ....

   ....

   ....

   b. Complete this summary equation for photosynthesis.

   .... + ... ⟶ .... + ...

2. **a.** Describe a chloroplast. ....

   ....

   b. Name the green pigment found in chloroplasts. ....

   c. Moira states that all plant cells contain chloroplasts. Explain why this statement is not correct.

   ....

   ....

   ....

   ....

3. Give three ways in which plants use the glucose they make during photosynthesis.

   ....

   ....

   ....

   ....

   ....

   ....

9.3 Evidence of photosynthesis: testing for starch

Thinking and working scientifically

1. Read the following paragraph and fill in the gaps with words from the box below. Each word may be used once, more than once, or not at all.

repeat	**variable**	**changed**	**reliable**	**measured**	**variables**

Scientific questions involve .... , one that can be .... to answer the question, and one that can be .... to see what effect that change has. There is usually more than one .... that can be measured. It is important to .... the measurements to check that the results are ....

2. The four sentences below describe how Salma investigated photosynthesis. The diagram shows the main stages of a scientific investigation. Write the letter of each sentence in the correct box.

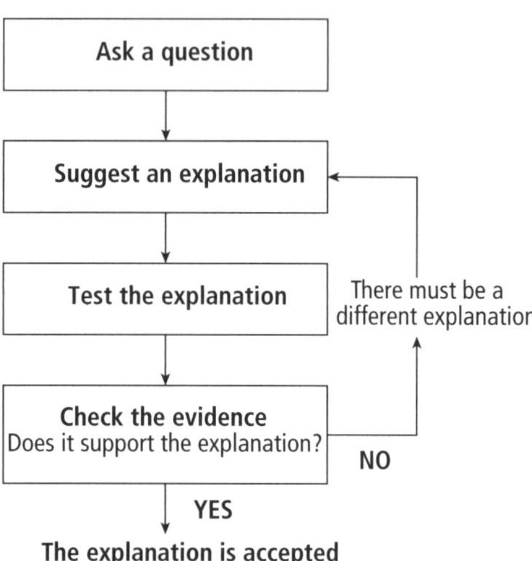

A Salma kept some plants in the light for several days and some plants in the dark. Then she tested leaves from both types of plants with iodine solution to see if they had made starch using the sugars made in photosynthesis.

B Salma wondered if plants really need light for photosynthesis.

C The results showed that only the plants kept in the light had starch in their leaves made from sugars produced in photosynthesis

D The parts of plants that are in the light are green and contain chlorophyll which traps light to use in photosynthesis. The parts of a plant that are in the dark all the time don't have any chlorophyll. So it seems likely that plants need light for photosynthesis.

3. a. Describe how you would test a leaf for the presence of starch and explain why each step is taken.

1 ....

....

2 ....

....

3 ....

....

4 ....

....

b. State the colour of iodine solution that indicates starch is present.

....

Extension

The iodine test for starch is often used on leaves to show that photosynthesis has taken place. Explain why the presence of starch shows us that the plant has been photosynthesising.

9.4 Evidence of photosynthesis: oxygen bubbles

Thinking and working scientifically

1. Rahul plans to use pondweed to answer this question:

   Does light intensity affect the volume of oxygen produced by pondweed each day?

   a. Using your scientific knowledge, predict what Rahul will see in his investigation and explain your prediction.

   ....

   ....

   ....

   ....

   b. Rahul makes five sets of apparatus. He places them all the same distance from the window and measures the mean volume of oxygen they produce in a day. Then he moves all the plants further away from the window and repeats his readings. His results are given in the table.

   Use these data to draw a line graph.

Distance from window (m)	Volume of oxygen collected (cm^3/day)
4.5	0
4.0	1
1.5	2
0.5	3
0.25	4

   c. State the conclusion that can be drawn from these results.

   ....

   ....

Extension

Sai wants to find out how the amount of carbon dioxide available affects the rate of photosynthesis. She has beakers of water containing differing concentrations of carbon dioxide, pondweed, beakers, measuring cylinders, funnels, and lamps.

a. Sketch and label a diagram of the apparatus Sai will set up. You need draw only one example.
b. Predict what results Sai will get from this investigation and give reasons for your answer.
c. Suggest one variable that Sai must control if her results are to be reliable.

9.5 The need for minerals

1. a. Colour the leaves below and add notes to describe how these mineral deficiencies affect the appearance of plant leaves.

Nitrate deficiency ....
....

Magnesium deficiency ....
....

b. Explain why nitrates are needed by a plant.

....

....

c. Explain why it is so important for plants to have plenty of magnesium.

....

....

2. Read the following paragraph and fill in the gaps with words from the box below. Each word may be used once, more than once, or not at all.

plant	**bacteria**	**soil**	**nitrogen**	**nitrates**
water	**root nodules**	**minerals**	**roots**	**legumes**

Plants get most of the .... they need from the .... The .... are dissolved in the .... in the .... and are taken into the .... through the .... The .... do not need to take in .... from the .... They have .... full of special .... which make .... from the .... in the air. .... use these .... themselves and add them to the soil, improving it for other plants to use.

Extension

Plan a demonstration to show a class of students which minerals are needed by plants to grow healthily. Describe how you would set up this demonstration, identifying the variables you will change, and the variable or variables you will observe and measure. Explain what you hope to be able to show your class of students.

9.6 The use of fertilisers

Science in context

1. Match the two parts of these sentences about the use of natural fertilisers:

A For thousands of years people have used	1 being cheap and improving the soil.
B Natural fertilisers include	2 natural fertilisers.
C Manure is made from	3 decomposed plant material.
D Compost is made from	4 nitrates are released slowly and in limited supply.
E Advantages of natural fertilisers include	5 decomposed animal droppings.
F Disadvantages of natural fertilisers include	6 manure and compost.

2. Answer these questions about artificial fertilisers.

a. State the approximate number of years that artificial fertilisers have been available for farmers to use. ....

b. Give two advantages of artificial fertilisers over natural fertilisers.

....

....

c. Give two disadvantages of artificial fertilisers compared to natural fertilisers.

....

....

3. The production and use of artificial fertilisers around the world is an example of science applied across industry. Write a paragraph describing each of these:

a. The scientific discovery of Fritz Haber that is the basis of the production of artificial fertilisers.

....

....

....

....

b. The work of Carl Bosch that allows farmers all over the world to use artificial fertilisers today.

....

....

....

....

9.7 Water and mineral transport in plants

1. Read the following paragraph and fill in the gaps with words from the box below. Each word may be used once, more than once, or not at all.

vacuoles	support	cell	wilts	water	absorb	photosynthesis

Plants need a continual supply of .... They use it for .... and it evaporates from their leaves. If plants lose more water than they can .... from the soil, their cell .... shrink. This makes each .... floppy. They cannot .... themselves so the plant ....

2. Label the arrows on the diagram to describe how water moves through a plant.

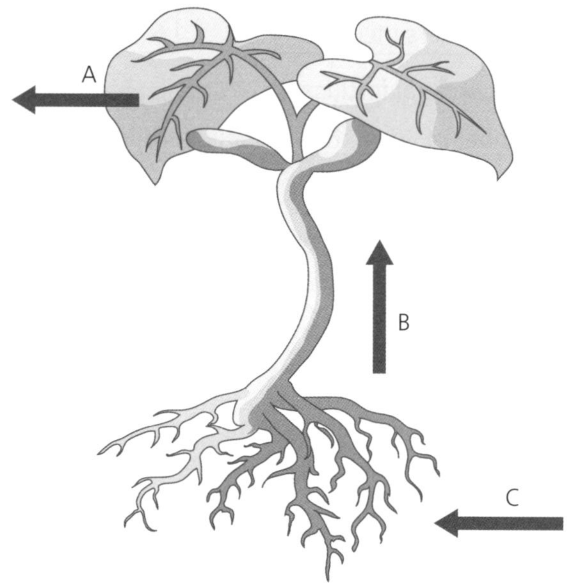

3. Draw lines to join each observation about transpiration to the correct explanation.

Observation
A Plants continually lose water
B Leaves have holes called stomata on the underside
C Root hairs give plant roots a large surface area
D Water flows from the roots to the leaves
E Plants lose less water at night

Explanation
1 to allow carbon dioxide to enter them.
2 because it evaporates from their leaves.
3 because their stomata close.
4 to help them absorb more water.
5 because it is pulled up the xylem tubes to replace the water lost by evaporation.

Extension

Students carried out an investigation. They smeared waterproof petroleum jelly over the underside of the leaves of a plant. They did not apply the jelly to the leaves of a similar plant. Both plants were placed by a sunny window and watered, and then left for several days. The plant without petroleum jelly on the leaves wilted first. Suggest an explanation for this observation.

Extension

1. a. Name the tissues labelled tissue A and tissue B in this diagram.

....

....

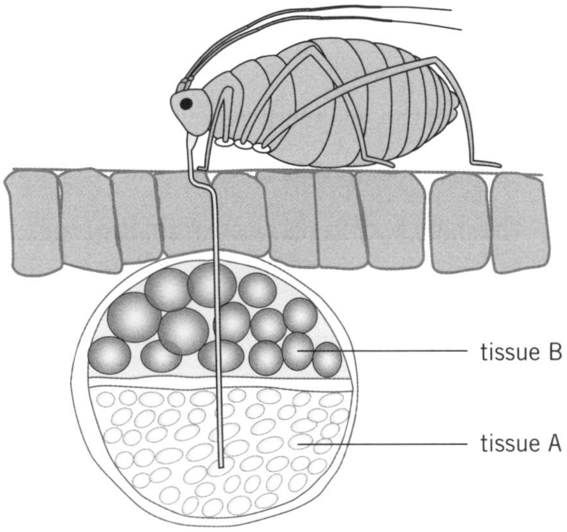

b. The insect shown in this diagram is an aphid. Explain why the aphid feeds from tissue A not tissue B.

....

....

....

....

c. Describe the main differences between tissue A and tissue B.

....

....

....

d. Use your scientific knowledge and understanding to describe the possible differences between tomato plants badly infected by aphids compared with tomato plants that are free of aphids and explain these observations.

....

....

....

....

....

....

10.1 What is excretion?

1. Use these words to complete the sentences below:

urea **excretion** **waste products** **toxic** **carbon dioxide**

The cells of your body carry out many reactions which result in ....

Some of these waste products are ....

The main poisonous waste product of aerobic respiration in your cells is ....

The main poisonous waste from the breakdown of proteins in your body is ....

The removal of these substances from your body is called ....

2. Complete this table by filling in the correct information about these two excretory organs.

Lungs	Kidneys

They excrete carbon dioxide.

The urea excreted is produced in the liver from the breakdown of excess proteins.

This is a pair of organs found in the chest, inside the ribcage.

The urea is removed from the body in the urine.

The carbon dioxide excreted is made in every cell during respiration.

This is a pair of organs found below the rib cage, against the back wall of the body.

The carbon dioxide is removed from the body when we breathe out.

They excrete urea.

3. Explain the difference between **excretion** and **egestion.**

Excretion:

....

....

Egestion:

....

....

10.2 The human excretory system

1. Name and label the parts of the human excretory (renal) system that carry out the following functions:
   - filters the blood and removes urea, some water and other substances your body doesn't need
   - carries urine from the kidney to the bladder
   - stores urine until it is full
   - carries urine from the bladder to outside the body.

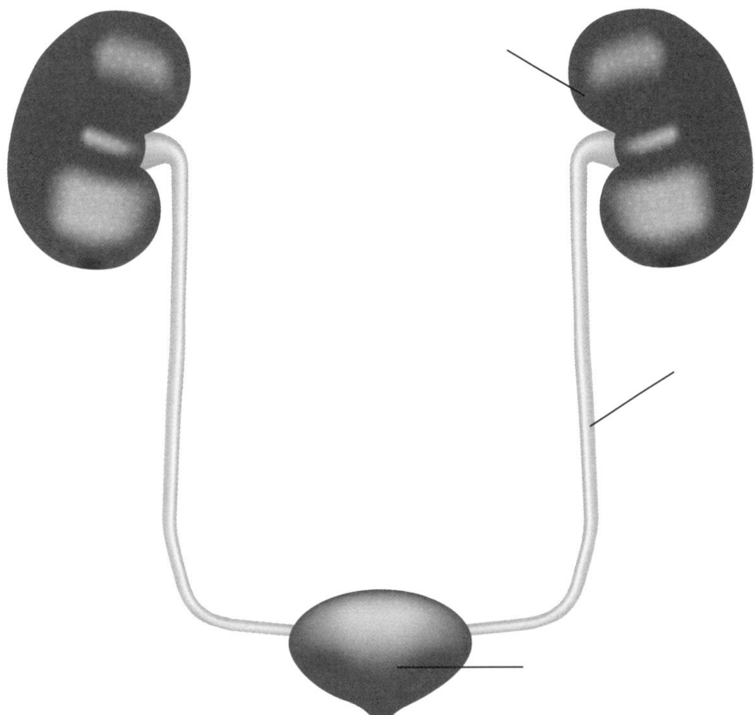

2. Read the following paragraph and fill in the gaps with words from the box below. Each word may be used once, more than once, or not at all.

urine toxic filter glucose proteins kidneys blood cells water urea blood

Your .... have a rich .... supply. They .... the ...., removing .... waste products such as .... but leaving .... and large .... behind. Useful substances like .... are filtered out of the blood but are returned to it. Some .... is also removed from the blood to form .... which carries the .... out of the body.

Extension

The body system shown in the diagram has several different names – the renal system, the urinary system and the excretory system. Some people have suggested that this body system should NOT be called the excretory system. Suggest reasons why the excretory system is not a good name to use for this organ system in the body.

10.3 Who made the best model?

Thinking and working scientifically

1.

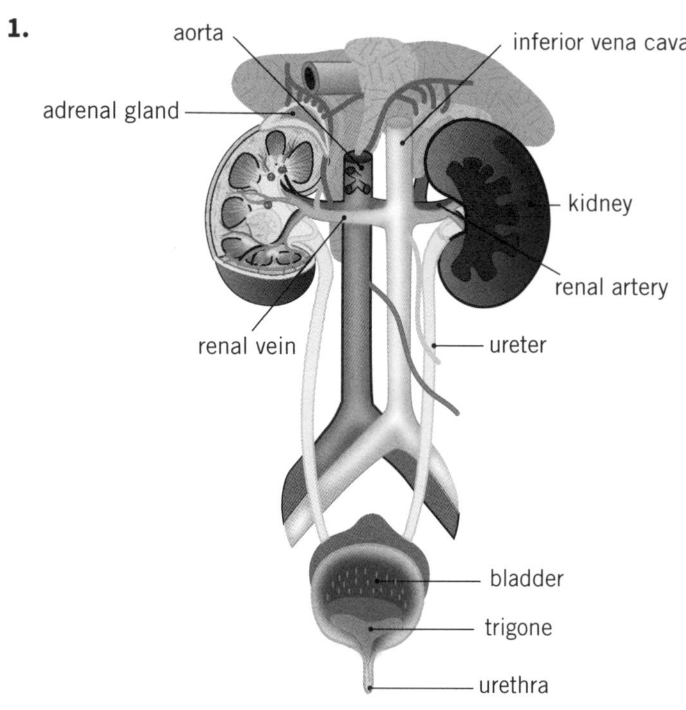

This illustration shows a plastic model of the human renal system.

a. Describe scientific models, explain what they are used for, and state the type of model shown in the diagram..

....

....

....

b. State one advantage and one disadvantage of using scientific models.

....

....

....

c. State one advantage and one disadvantage of using a plastic model of the renal system like the one shown here.

....

....

....

d. Describe an alternative way of modelling the human kidney. Give one advantage and one disadvantage of your suggested model over the plastic model shown.

....

....

10.4 When kidneys go wrong

Science in context

1. Use the words below to complete the following paragraph.

urea	**rejection**	**live**	**toxic waste**	**two**
transplant	**urine**	**kidney**	**products**	**deceased**

If you receive an organ .... , you need to take special drugs to avoid .... The organ most often transplanted is the .... We have .... kidneys. They remove .... like .... from the blood and make .... Most kidney transplants are from .... donors but we only need one kidney to stay healthy, so a .... donor can give one away.

2. Use lines to match each word to the correct definition.

Word
A transplant
B reject
C urine
D kidney

Definition
1 the organ that removes urea from your blood and makes urine
2 a waste product made by your kidneys and stored in your bladder
3 move an organ from one person to another
4 fail to accept a transplanted organ

3. Some of these sentences are true and some are false. Write T after the true statements and F after the false ones. Rewrite the false statements so they are correct.

a. An organ can be kept alive for a short time outside the body. ....

b. Organs can only be transplanted from close relatives. ....

c. After a transplant, special medicines need to be taken for a few weeks to prevent rejection. ....

d. Organs grown from a patient's own tissues or from an identical twin are never rejected. ....

....

....

Extension

Read the information in the box. Then answer the questions below.

Lives can be extended and greatly improved by replacing faulty organs with healthy organs in an organ transplant. Waiting lists for organ transplants are often very long, and they are growing longer all the time. Scientists want to solve this problem. They are trying to grow new organs from people's own tissues. They can already make simple body parts like windpipes and bladders. Organs like hearts and lungs will be more difficult to make.

a. Suggest why more transplants will be needed in future.

b. Explain why a heart will be harder to grow than a bladder.

10.5 Kidneys work everywhere

Extension

1. The data in the table shows the mean daily urine output of a number of species of mammals.

Species	Mean daily urine output (cm^3/kg)
horse	10
cattle	31
sheep or goat	25
cat	30

a. Use these data to draw a bar chart. Leave space to add two more bars to your graph later.

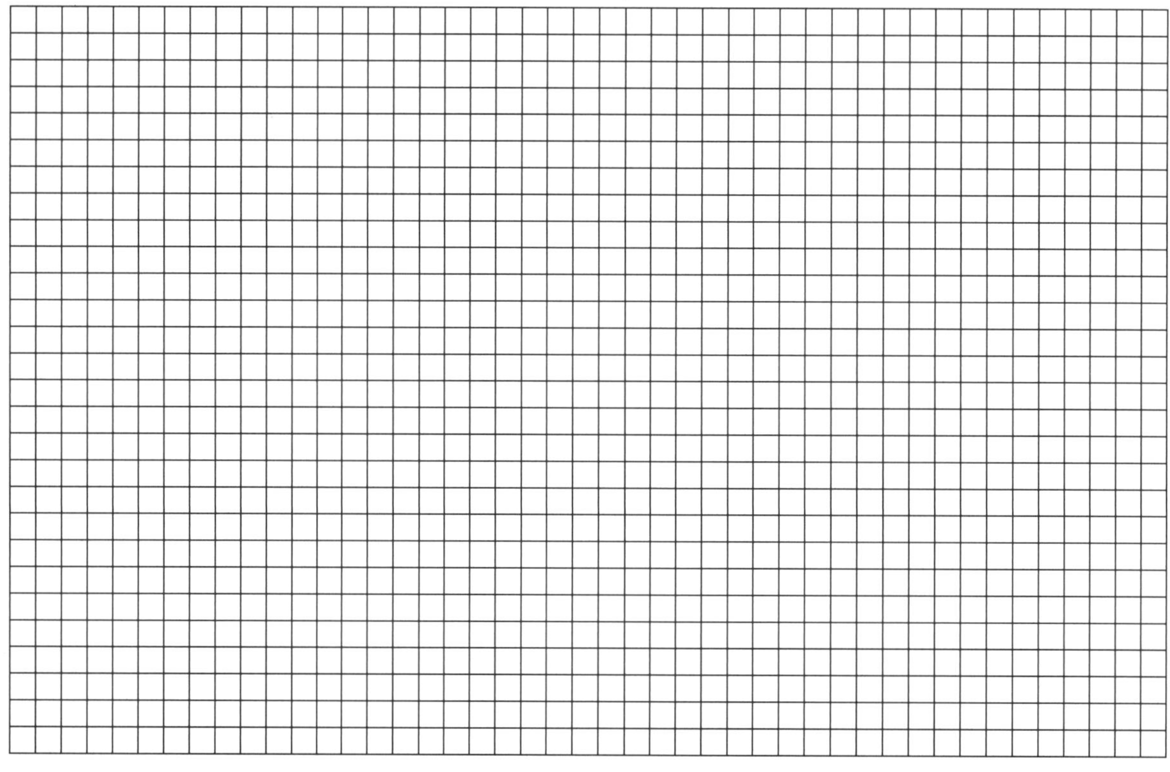

b. i. Animals that live in deserts like the kangaroo rat have many challenges but they still need to excrete urea. Draw and label a bar on the graph to predict the mean daily urine output of a kangaroo rat in cm^3/kg.

ii. Explain your prediction.

....

....

c. i. Mammals like beavers that live in fresh water are never short of water. Draw and label a bar on the graph to predict the mean daily urine output of a beaver in cm^3/kg.

ii. Explain your prediction.

....

....

11.1 Reproduction: a characteristic of life

1. Match these terms to their definitions:

Term
A genetic material
B DNA
C chromosomes
D gametes

Definition
1 long threads of DNA formed before a cell divides
2 special reproductive cells that contain half the number of chromosomes seen in normal body cells
3 the material passed from parents to their offspring
4 the genetic material in cells which contains the instructions for new life

2. Fill in the gaps in the following paragraph with words from the box below. Each word may be used once, more than once, or not at all.

sexual similar two offspring DNA different asexual

There are .... main types of reproduction. In .... reproduction, only one parent organism is involved. The .... produced are identical to their parent. In .... reproduction, ... parents are involved. The .... get some .... from each parent, so they are .... to, but also .... from, both parents.

3. Complete the table below to show the number of chromosomes found in the body cells and the gametes of different species of animals and plants.

Species	Number of chromosomes in body cells	Number of chromosomes in gametes
human	46	
elephant		28
coconut palm	32	
boa constrictor		18
torch ginger	48	
tortoise		26

Extension

A plant has 20 chromosomes in each of its body cells. It reproduces both asexually and sexually. All of the new plants formed have 20 chromosomes in their body cells.

a. State the main difference between the plants produced by asexual reproduction and the plants produced by sexual reproduction.

b. Explain how this difference comes about.

11.2 Fertilisation: new life begins

1. Read the following paragraph and fill in the gaps with words from the box below. Each word may be used once, more than once, or not at all.

fertilisation	sperm	DNA	egg cell
chromosomes	ovary	egg	nucleus

A mature .... is released from the .... of a female mammal. Once millions of .... are released inside the body of the female by the male, they swim towards the .... A new life begins when a .... nucleus fuses with an egg cell .... This is .... The fertilised egg has a new combination of .... because half of the .... are from the female parent and half are from the male.

2. a. State what is meant by a gamete and describe how a gamete differs from a normal body cell.

....

....

b. Label the male and female sex cells shown below and explain how their structures are suited to their functions in the body.

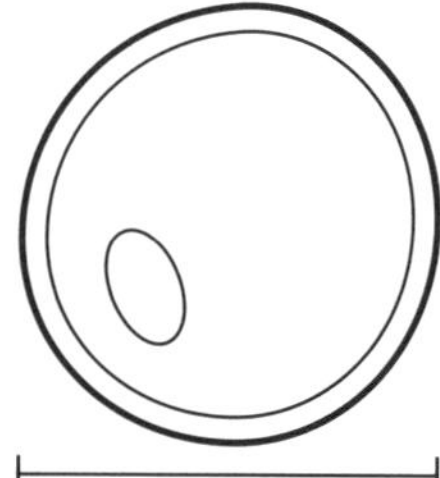

3. Add a label to each of the diagrams below to describe how two sex cells join to make a new life.

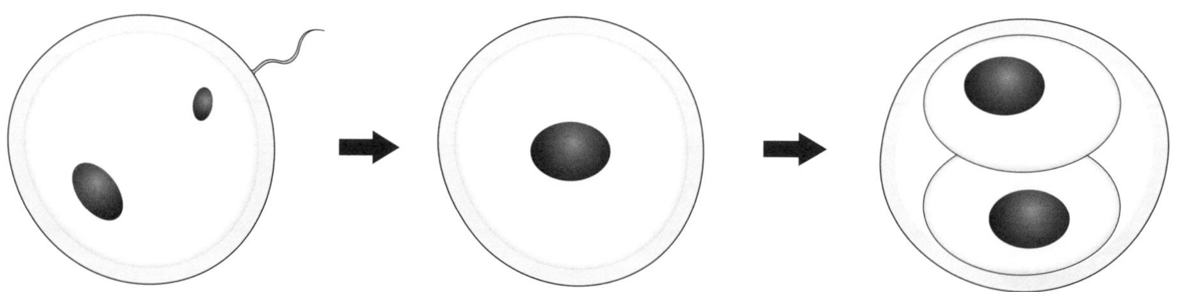

Extension

Briefly explain how the development of a baby from a fertilised egg is controlled.

11.3 Boy or girl? Sex inheritance in humans

1. Decide whether these sentences are true for 'genes', 'chromosomes' or 'both'.
   a. Found in the nucleus ....
   b. Inherited from our parents ....
   c. Visible just before a cell divides ....
   d. Made from giant molecules called DNA ....
   e. Make from special sections of DNA ....
   f. There are thousands in each cell ....
   g. There are 23 pairs in each human cell ....
2. This diagram shows a human cell. 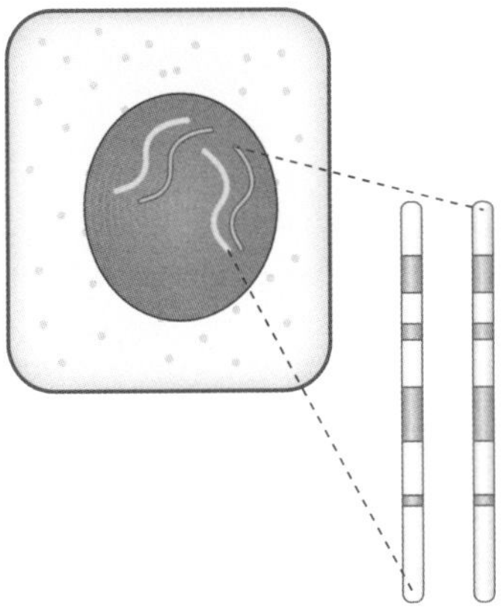
   a. Add labels to the diagram to the cell nucleus, chromosomes, and genes.
   b. Describe two differences between the diagram and a real cell.

      ....

      ....

      ....

      ....

   c. Explain how to tell whether a human cell is from a boy or a girl.

      ....

      ....

      ....

      ....

3. Label the diagram and use it to explain why fertilised eggs are equally likely to be male or female.

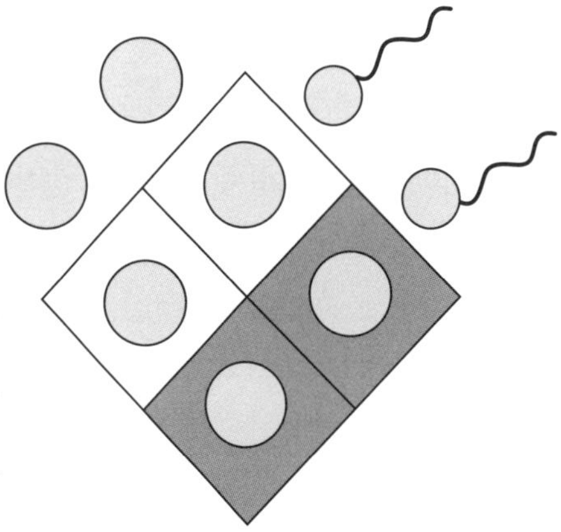

   ....

   ....

   ....

   ....

   ....

   ....

   ....

   ....

   ....

   ....

11.4 Variation between individuals

1. Read statements **a–g** below. Write T after the ones that are true and F after the ones that are false. Write the correct version of false statements on the lines below.
   - **a.** Your genes control the activities of your cells. ....
   - **b.** We inherit genes from both parents so we inherit some of their features. ....
   - **c.** Twins always have identical genes. ....
   - **d.** Identical twins develop differences when their environments differ. ....
   - **e.** A poor environment can prevent your genes from keeping you healthy. ....
   - **f.** Every cell in your body contains identical genes. ....
   - **g.** Cells specialise by switching on all their genes. ....

   ....

   ....

   ....

   ....

2. **a.** The table below shows the distribution of blood groups in India. Use the data to plot a frequency chart on a piece of graph paper.

Blood group	Number of people (%)
O	39
A	23
B	32
AB	6

   b. Blood groups are an example of inherited variation. Explain this statement.

   ....

   ....

   ....

   c. Height is another characteristic that varies between individuals. Explain why the average height of students is increasing in many parts of the world.

   ....

   ....

   ....

Extension

Scientists want to know whether your genes or your environment have the biggest influence on your behaviour. To answer this question, they often compare identical twins who were separated at birth with identical twins who grew up together.

a. Suggest why identical twins separated at birth are ideal for these studies.

b. What results would you expect if inherited variation turned out to be more important than environmental variation in determining your behaviour?

11.5 The development of a fetus

1. Put the following sentences in order to show how a fetus develops.
   Start with statement G.

   The correct order is:..

   A The embryo becomes a hollow ball of cells.

   B The fetus has most of its organ systems but they need time to grow and develop.

   C The fetus is mature enough to survive outside the uterus.

   D The fertilised egg cell begins to divide to form an embryo.

   E The embryo's cells form tissues and organs and it becomes a fetus.

   F The embryo implants in the wall of the uterus and develops a placenta.

   G A sperm cell nucleus fuses with an egg cell nucleus during fertilisation.

2. Here is a developing human embryo in the uterus.

   a. Label the diagram.

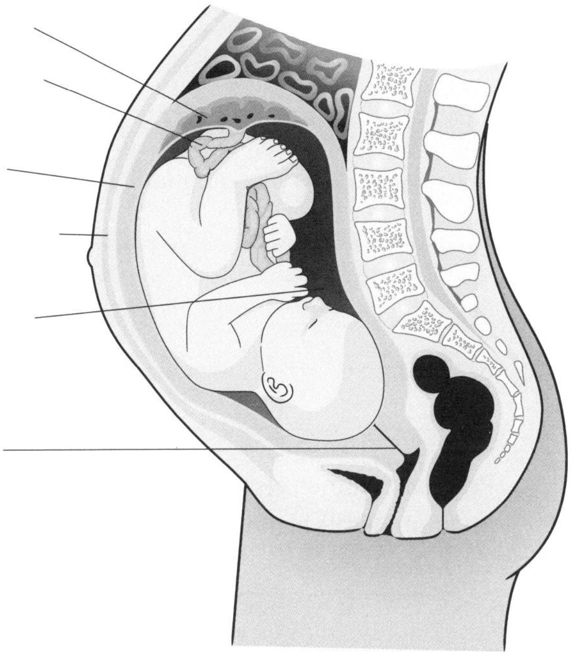

   b. Describe the function of the umbilical cord.

   ....

   ....

   c. Explain why the fluid in the fluid sac around the fetus is important.

   ....

   ....

   d. Explain what happens when the blood from the fetus flows through the placenta.

   ....

   ....

   ....

   ....

11.6 Health of the mother, health of the child

1. The diagram below represents the thin barrier between the blood of the fetus and the blood of the mother in the placenta.

a. Add labels to these arrows to show which way the following molecules move:

oxygen carbon dioxide glucose urea.

b. Use this diagram to help you explain why doctors are so careful about the medicines they give to pregnant women.

....

....

....

....

2. a. State how a fetus gets the nutrients it needs from its mother.

....

....

b. Give two reasons why pregnant women need to eat plenty of iron-rich foods.

....

....

....

....

c. Pregnant women also need to eat plenty of calcium-rich foods in their diets. Use what you know about minerals in the diet to suggest why this is important.

....

....

....

d. Suggest one other nutrient that will be important in the diet of a pregnant woman and explain your choice.

....

....

Extension

Discuss three of the problems seen in babies born to mothers who have taken drugs during their pregnancy.

11.7 Understanding science, saving lives

Science in context

1. Some babies have problems with their sight or their hearing. This can have big effects on the lives of those children. They may need a lot of help and support. Scientists are finding some of the causes of sight and hearing problems. We may be able to prevent some of them.

   A hospital compared babies with low and healthy birth weights. The results are shown below.

   a. Plot these results as a bar chart.

Birth weight (kg)	Poor vision (%)	Hearing problem (%)
less than 1	50	46
more than 3	1	1

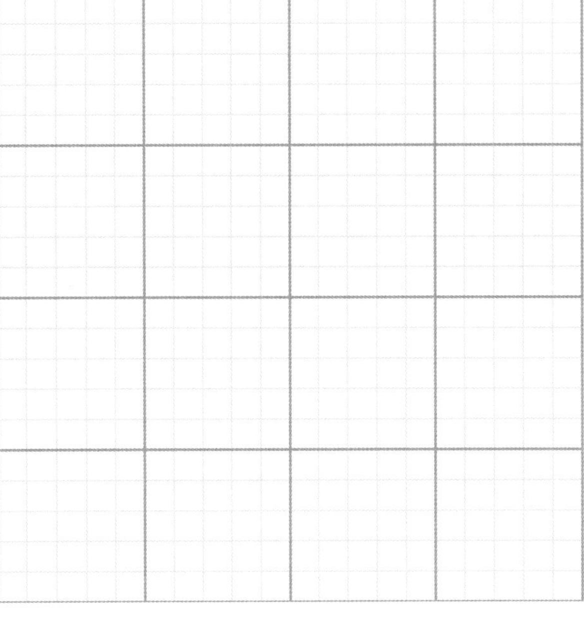

   b. Use these to help you explain why it is so important for pregnant women to have enough to eat in the final three months of pregnancy.

   ....

2. Some babies are born small because they are born too early. We do not always understand why this happens. Some are born small because their mothers have not had enough to eat during their pregnancy. Some are born small because their mothers have used drugs such as caffeine during their pregnancy causing the babies to be born small.

   Discuss how these issues could be helped by scientific understanding on the part of the mother and father and their community.

   ....

   ....

   ....

   ....

   ....

   ....

11.8 Smoking and pregnancy: the evidence

Thinking and working scientifically

Doctors studied 189 new mothers in a US hospital during the 1980s. They recorded the smoking habits of the mothers and the birth weight of their children. Here are some of the data.

a. The medical definition of a low birth weight is one below 2500 g. Calculate the percentage of non-smokers and smokers who had low birth weight babies. Show your working.

Birth weight range of babies (g)	Non-smokers	Smokers
500–999	0	1
1000–1499	3	1
1500–1999	8	6
2000–2499	18	22
2500–2999	22	16
3000–3499	29	17
3500–3999	27	10
4000–4499	6	1
4500–4999	2	0

....

....

....

....

....

b. Plot these data on a bar chart, one under the other on the same axes. Use birth weight (g) along the horizontal axis and numbers of babies on the vertical axis.

c. Evaluate these findings and compare them to the data you have studied looking at the link between smoking and birth weight in your text book.

....

....

....

d. Suggest a reason why mothers who smoke have an increased risk of a low birth weight baby.

....

....

12.1 The carbon cycle

1. The diagram shows the carbon cycle in nature.

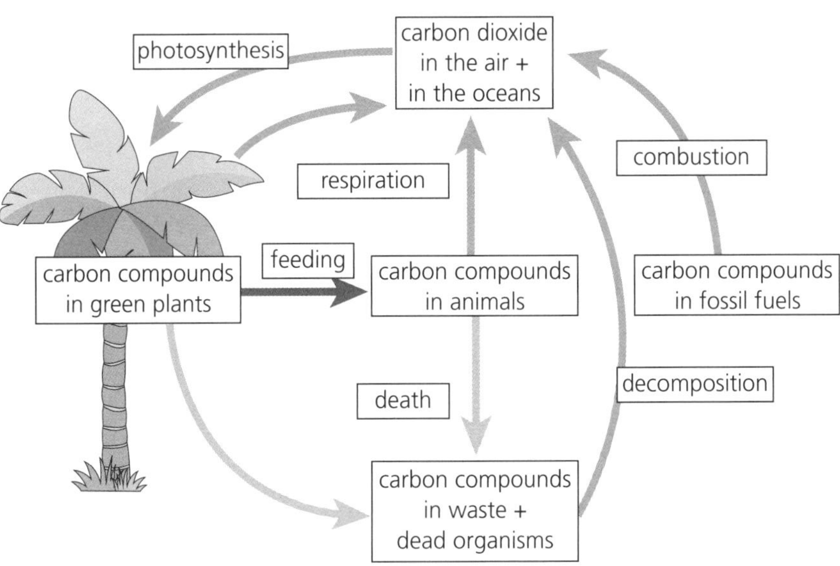

Look at the different processes involved in the carbon cycle and explain what happens to the carbon in each one:

Photosynthesis:..

....

....

Feeding:..

....

....

Respiration:..

....

Decomposition:..

....

....

Combustion:..

....

....

Extension

Discuss why it is so important that students learn about the carbon cycle in the 21st century.

12.2 People and the carbon cycle

Science in context

1. Read the following paragraph and fill in the gaps with words from the box below. Each word may be used once, more than once, or not at all.

warm	greenhouse gases	carbon dioxide	temperature	Earth	Sun
oxygen	organisms	live	water	energy	carbon

On the surface of the .... we have the ...., ..., and .... needed for life. The .... at the surface is ideal for .... to .... and grow. This is maintained by the presence of .... such as .... that trap enough of the .... from the .... to keep the surface of the Earth .... but not too hot.

2. Over the last two centuries human activities have increased the levels of carbon dioxide in the atmosphere.

   a. Describe and explain the effect of a rise in the concentration of carbon dioxide in the atmosphere.

   ....

   ....

   ....

   ....

   b. Give three examples of ways in which humans are affecting the levels of carbon dioxide in the atmosphere and explain how each activity has its affect.

   1 ....

   ....

   2 ....

   ....

   3 ....

   ....

Extension

Different parts of the world produce different amounts of carbon dioxide. Discuss some of the factors that affect the carbon dioxide production of a particular region.

12.3 Historical impacts of climate change

1. Match the start and end of the following sentences:

A Scientists have gathered evidence showing
B The Earth has had periods of
C When the temperature of the Earth was very low,
D When the temperature of the Earth was hot,
E Global warming events have always been linked to

1 very high and very low temperatures.
2 there were huge areas of deserts without water.
3 a rise in carbon dioxide in the atmosphere.
4 that the climate has changed over millions of years.
5 the surface was covered in ice and snow.

2. The cold periods in the history of the Earth are known as Ice Ages. Describe four possible reasons why the surface of the Earth becomes so cold:

A. ....

B. ....

C. ....

D. ....

3. Every country has been affected by global warming over the history of the Earth. Complete the table below to summarise the main historical impacts of global warming.

Observed change	Explanation
sea levels rising	
	Global warming and climate change are linked to increases in very heavy rainfall which lasts a long time and causes this problem.
drought	
extreme weather events	

12.4 Predicting the future

1. In 1896 CE, a scientist called Svante Arrhenius made a prediction: doubling the amount of carbon dioxide in the atmosphere would make it 5 °C warmer at the surface of the Earth.

   a. State Arrhenius's prediction about carbon dioxide levels and global temperature. ....

   ....

   b. Look at these graphs. Describe what they show and discuss whether this supports Arrhenius's prediction. ....

   ....

   ....

   ....

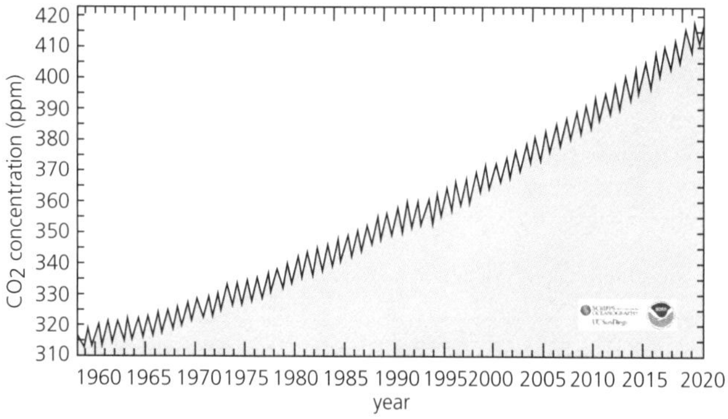

Global surface temperature records

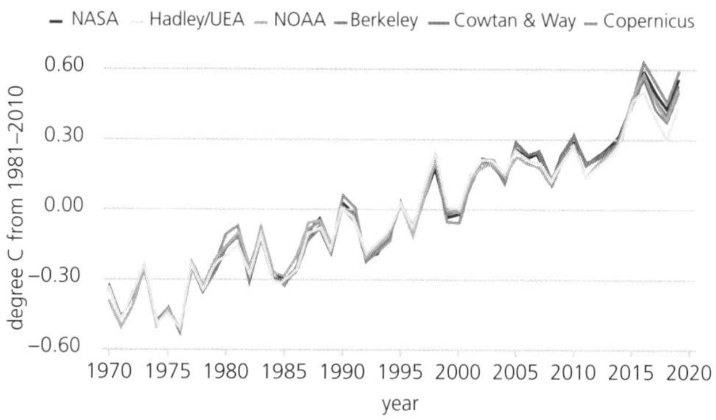

Annual global mean surface temperatures

2. The IPCC predict that, by the middle of the 21st century, global temperatures will increase by another 1.3–1.8 °C.

   If global temperatures increase as the IPCC predicts, suggest three possible effects of this global warming and explain the problems they will bring to people living in the affected areas.

   ....

   ....

   ....

   ....

   ....

   ....

Extension

The IPCC prediction of global temperature increases covers a range of 1.3–1.8 °C. Discuss why they give a range of temperatures and what sort of factors will affect which temperature increase is actually observed.

12.5 Evaluating the evidence for climate change

Extension

1. a. Scientists use different types of evidence to help them understand climate change. Complete the table to describe what the evidence shows.

Measurements taken	What the evidence shows
past temperature readings from weather stations all over the world	
the thickness of tree rings	
the amount of carbon dioxide trapped in layers of ice	

b. Explain how this evidence is useful to scientists today.

....

....

....

2. Scientists analyse ice cores to get carbon dioxide data from thousands of years ago.

a. Describe what scientists analyse from these ice cores.

....

....

b. Explain how this data is useful to scientists today.

....

....

....

c. Describe the changes seen in the levels of carbon dioxide in the atmosphere over the last two centuries and suggest why many scientists worry about what they are observing.

....

....

....

13.1 Variation in animals and plants

1. Read the following paragraph and fill in the gaps with words from the box below. Each word may be used once, more than once, or not at all.

offspring	**environmental**	**characteristics**	**inherited**
environment	**genes**	**genetic**	**combination**

Genetic or .... variation is carried in the .... passed on from parents to their .... The other main type is .... variation which is a result of the .... where an organism lives. Many........................... are the result of a .... of .... and .... factors.

2. a. Explain why organisms from two different species usually look very different to each other.

....

....

b. Explain why organisms of the same species all look similar.

....

....

c. Explain why organisms of the same species do not look identical to each other.

....

....

3. Two families of lions live in the same part of the Serengeti area of Africa. The lion cubs in one family have longer legs and are also heavier than the cubs in the other family. Suggest reasons for these variations between animals of the same species.

....

....

....

....

13.2 Natural selection in action

1. Put these sentences in the correct order to describe the process of natural selection. Number the sentences 1–5.

	This process is repeated many times until these characteristics become more common in the population. Over a long period of time, it may lead to the development of a new species.
	This is natural selection in action.
	Each individual inherits genetic variation from their parents and is different from all other members of their species.
	The successful individuals are most likely to reproduce and pass on their useful characteristics.
	The individuals with characteristics that give them an advantage in their particular environment are the ones most likely to survive to become adults.

2. This diagram shows the bones in the front limbs of four different mammals. The evidence shows that all mammals had a common ancestor millions of years ago, but their limbs are different as a result of natural selection for different environments.

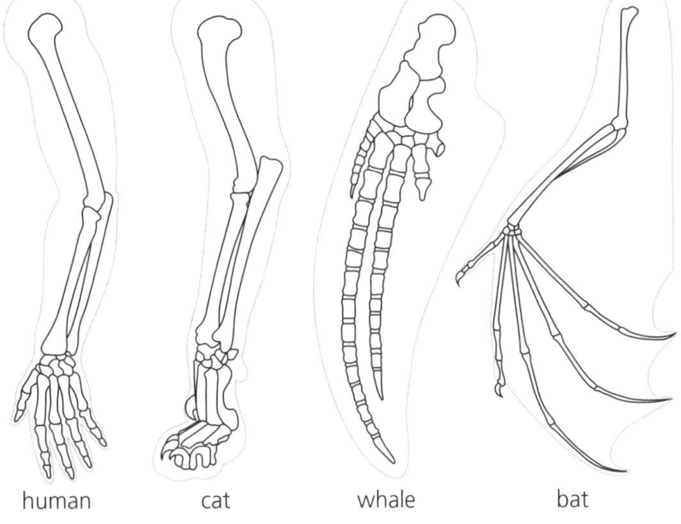

a. Look carefully at the limbs of these four mammals. Suggest how they give us evidence that all mammals share a common ancestor.

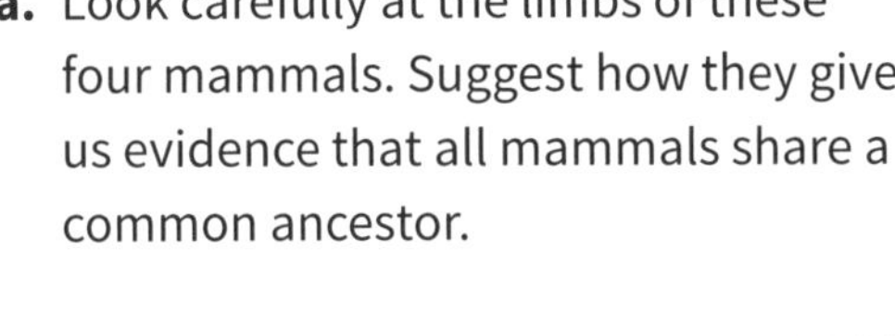

....

....

....

b. Natural selection involves the survival of the best adapted offspring for a particular environment. Describe two adaptations of the front limb of the bat which make it well adapted for flying.

....

....

c. Describe two adaptations of the front limb of the whale that make it useful for swimming.

....

....

Extension

Antibiotics are medicines used to cure bacterial diseases by killing the bacteria that cause the illness or by stopping them growing. Some bacteria are becoming resistant to these antibiotics, which means they are no longer affected by the medicines.

a. Explain how antibiotic resistance might develop in a population of bacteria.

b. Discuss why the development of antibiotic resistance in bacteria is such a problem for people.

13.3 Environmental change and natural selection

1. Read the following paragraph and fill in the gaps from the box below. Words may be used once, more than once or not at all:

offspring	common	generation	characteristics
species	population	natural selection	survive

Members of a .... are not all the same. Some have .... that make them more likely to .... and raise .... They pass on these useful .... to the next .... and, over many generations, they become more .... in the .... This process is known as ....

2. Scientists have evidence that polar bears that live in the Arctic have developed from brown bears over thousands of years by a process of natural selection. Brown bears moved northwards in a period when the climate of the Earth was relatively warm. When the next Ice Age happened, the areas where some of the bears lived became very cold, with a lot of snow and ice. Answer the questions below on the process which resulted in two species – brown bears and polar bears – that we still see today.

   a. Occasionally, brown bear cubs were born with white fur. Suggest how this white fur would help bears survive in a cold, icy environment.

   ....

   ....

   b. Explain how the genes for white fur might become more common over many generations of bears living in or near the Arctic.

   ....

   ....

   ....

   c. Explain why we still have both brown bears and polar bears.

   ....

   ....

   ....

   ....

   d. Some people predict that polar bears will eventually die out. Explain why and how polar bears might be lost as a species.

   ....

   ....

   ....

   ....

13.4 Extinction!

1. Read the following paragraph and fill in the gaps from the box below. Words may be used once, more than once, or not at all.

extinct habitats biomass plant species tropical biodiversity destroyed

Many different .... species grow in .... rainforests. These produce a vast quantity of .... each year and provide a wide variety of different .... When they are cleared to grow crops, ... falls quickly because most of the .... are .... Species which are found nowhere else become ....

2. Scientists have observed the rate at which different species have gone extinct over time. They use these observations, along with others, to work out the causes of animal extinctions and predict what is likely to happen in the future. Here is some of their data:

Vertebrate group	Threatened species (% of total in 2000)	Threatened species (% of total in 2020)
fish	3	20
amphibians	3	41
reptiles	4	34
birds	12	14
mammals	24	26

a. Plot a bar chart to display these data on the graph paper below.

b. Discuss which groups have seen the greatest increases in the threat of extinction and which have seen the least.

....

....

....

c. Suggest three reasons for the increased risks of extinction shown in mammals since 2000.

....

....

....

Extension

Scientists around the world are conserving plant species by saving their seeds in cool dry conditions where they will last for hundreds of years. Explain why it is harder to conserve animals and prevent them from going extinct than it is to conserve plants in this way.

13.5 Investigating the peppered moth: past and present

Thinking and working scientifically

1. Match the two parts of these sentences so they make sense:

A The species name of peppered moths is
B Peppered moths are found from China and Russia to
C The colour of peppered moths varies from
D Well-known studies on these moths were carried out by
E The changes in peppered moth populations is an example of

1 natural selection in action.
2 almost white to almost black.
3 Professor Kettlewell in the UK.
4 Europe and North America.
5 *Biston betularia*.

2. The graph below shows the changes in the dominant colour of peppered moths from 1800 to the 2000s.

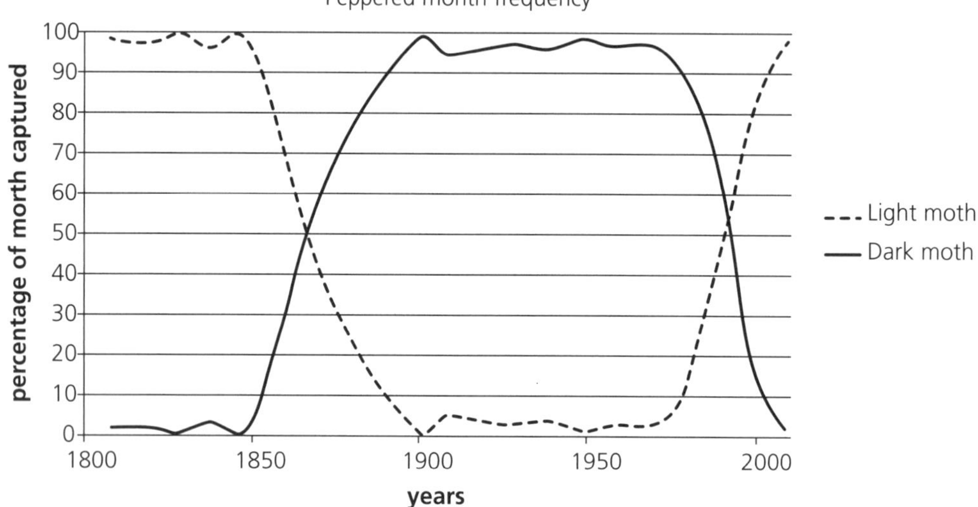

a. Describe in detail what this data tells you.

....

....

....

....

....

....

b. Suggest explanations for the trends shown in the data at different times.

....

....

....

....

....

....

13.6 What can we do?

Science in context

1. The bar chart below shows the increase in the use of hydroelectric power in several countries between 2000 and 2020.

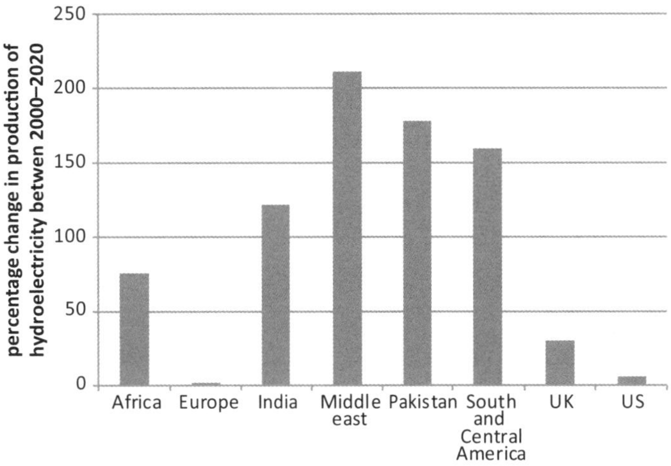

a. State the two countries or regions with the biggest percentage increase in the use of hydroelectric power and the two countries or regions with the lowest increase in the use of hydroelectric power.

....

....

....

....

percentage change in production of hydroelectricity betwen 2000–2020

250
200
150
100
50
0

Africa Europe India Middle east Pakistan South and Central America UK US

b. i. The fact that a country or region has a big increase in the percentage production of hydroelectric power over time does not always mean it is generating a lot of electricity using hydropower. Explain this statement.

....

....

....

ii. The fact that a country or region has a small increase in the percentage production of hydroelectric power over time does not always mean it is not generating much electricity using hydropower.Explain this statement.

....

....

....

c. i. Describe and explain one potential benefit of using hydroelectric power to generate electricity.

....

....

ii. Describe and explain one potential disadvantage of using hydroelectric power to generate electricity.

....

....

1.1 What is life?

1. A3; B5; C6; D7; E1; F4; G2
2. The statements that apply to all living things are: a, b, d, e, f, g.
3. Any three from: they move by extending their stems and roots; they use respiration to release the energy they need to stay alive; they sense things in their surroundings, such as light; they increase in size during their lifetime; they produce offspring; they remove waste products from their bodies; they make nutrients.

Extension

- **a.** Any two from: respiration would remove oxygen from the surroundings; add carbon dioxide to the surroundings; raise the temperature as heat transferred to the surroundings.
- **b.** Any reasonable suggestion that shows your understanding of the biological processes involved in living organisms – observe for signs of movement, change conditions, e.g. temperature or light and see if there is any response, etc.

1.2 Investigating living organisms: yeast

1. The missing words are: change, question, measure, effect, controlled, same.
2. **a.** The nutrient added.
   - **b.** The diameter of the balloon.
   - **c.** Any three from: quantity of yeast; mass of nutrient added; temperature of the water; the size/type of balloon; the size/shape of the flask.
   - **d.** Carbon dioxide.
   - **e.** Respiration (fermentation).
   - **f.** Glucose.
   - **g.** In a fair test, we change one **variable** to find out what effect it has. To make it a fair test, we must keep all the other variables the same.

Extension

- **a.** Any two from: different groups did things slightly differently; some groups made mistakes in their measurements; the balloons may have been different sizes or have stretched more or less easily; may have used different yeast cultures.
- **b.** All living organisms vary so no investigation using living organisms can be a completely fair test, because there will always also be unknown variables in the living organisms themselves.

1.3 Classification and species

1. The missing words are: characteristics, species, Latin, world, infertile, hybrids.
2. **a.** **Same species:** none; **Related species:** *Equus ferus; Equus africanus*; **Very different species**: *Tetracerus quadricornis; Syncerus caffer.*
   - **b.** Similar species are given the same first name in their two part Latin names. The first part of the species name of a zebra is *Equus* so other species with the same first name will be similar/ also be members of the horse family.
3. **a.** F; **b.** T; **c.** F; **d.** T; **e.** F
   - **a.** Similar species share the same first Latin name.
   - **c.** Members of the same species don't always look similar.
   - **e.** Members of different species usually have infertile offspring if they breed.

Extension

- **a.** Dzos have not bred to form large herds because they are infertile hybrids.
- **b.** Yaks and dzos could be distinguished using breeding experiments – only yaks would produce fertile offspring.

1.4 Classifying invertebrate animals

1. The missing words are: differences, groups, classification, backbones, characteristics, species.
2. Top row: **molluscs** – muscular bodies/often have a shell; **cnidarians** –tentacles/stinging cells; **annelids** – segmented bodies; **flatworms** – flat bodies.

   Bottom row: **echinoderms** – spiny skin; **arthropods** – jointed legs; **nematodes** – long thin bodies.
3. **Arachnids** – 8 legs/many spin webs; **insects** – 3 segments to the body/6 legs; **crustaceans** –at least 10 legs; **myriapods** – long segmented body/1 or 2 pairs of legs on each segment.

Extension

a. It is an arthropod because it has jointed legs and a hard outer covering; it is an insect because it has 6 legs. It also has three body segments – a head, thorax and abdomen.

b. i. Secondary source of information is information that has been collected by other people, not the result of your own observations.

ii. Disadvantage: you don't know how reliable or accurate the observations are. Advantage: you can get information on organisms from all over the world/you can get observations from equipment you don't have yourself.

1.5 Simple keys

1. **a.** B

   b. It has hollow fangs, a heart-shaped head and 18–22 V-shaped stripes.

   c. A is another *Bitis* species – has hollow fangs and a heart shaped head – but a different pattern; C is another type of venomous viper because it has hollow fangs BUT it does not have a heart-shaped head so it is NOT a type of African adder.

Extension

Advantages of diagrams: can highlight features from key; safe; can access organisms from anywhere. Disadvantages: can't see many features, can't see behaviour.

Advantages of photos: more realistic than drawings; safe; can access organisms from anywhere. Disadvantages: don't show all the features; can't see behaviour.

Advantages of live organisms: can see from different angles; can observe over time; can see/hear behaviours. Disadvantages: may be toxic/fierce/dangerous; may harm the organisms; expensive; can't easily get live organisms from many different parts of the world together in one place.

1.6 Classifying vertebrates

1. **a.** They all have a backbone.

   b. **Fish**: Three from: scales covering body; gills for breathing in water; fins and a tail; lay eggs in water to reproduce; body temperature varies with their surroundings.

   Amphibians: Three from moist skin; four legs; breathe through skin and lungs when adult; lay eggs in water to reproduce; body temperature varies with their surroundings.

   Reptiles: Three from dry scaly skin; lay eggs with leathery skins on land; have four legs (except snakes); use lungs to breathe air; body temperature varies with their surroundings.

   Birds: have feathers and beaks; lay eggs with hard shells on land; two legs and two wings; breathe air using lungs; control their own body temperature.

   Mammals: have fur; give birth to live young/make milk to feed them; four limbs; breathe air using lungs; control their own body temperature.

2. **a.** Mammals.
   b. Mammals have fur or hair and feed their young on milk.
   c. Echidnas lay eggs instead of giving birth to live young.

Extension

a. *Archaeopteryx* is difficult to classify because it has features from more than one group.

b. It shares features with birds (feathers and a beak) and reptiles (teeth and a bony tail).

1.7 Classifying plants

1. **a.**

Type of plant	Liverworts and mosses	Ferns	Conifers	Flowering plants
Roots and veins?	no	yes	yes	yes
Spores or seeds?	spores	spores	seeds	seeds
Cones, flowers, or neither?	neither	neither	cones	flowers

b.

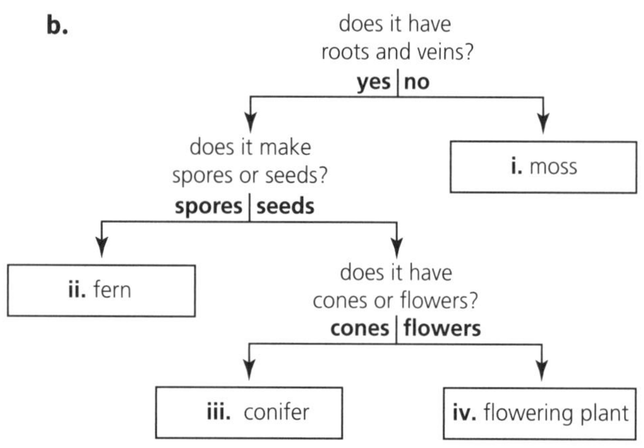

2. A. conifer: needle-like leaves and cones; B. flowering plant: leaves, roots and flowers; C. fern: small root-like organs, feathery leaves, no flowers or cones, relatively big; D. moss: small, no proper roots or leaves, no flowers or cones.

1.8 Making your own identification keys

1. Accept any unambiguous and effective key, e.g.

1	Has long legs	See 2
	Does not have long legs	See 3
2	Has a long thin neck	ostrich
	Has a short neck	stilt bird
3	Has a long thin beak	See 4
	Has a short beak	See 5
4	Flies	humming bird
	Does not fly	kiwi
5	Has webbed feet	puffin
	Has feet with claws	parrot

1.9 Are viruses living?

1. The missing words are: everywhere, bacteria, invade, living organisms, virus-making factories, diseases, parasites.
2. **a.** Protein coat, genetic material.
   b. They are very small so it is very hard to see them. They only reproduce in living organisms so it is difficult to grow them in the laboratory.
   c. Any two human viral diseases, e.g. colds, flu, COVID-19, measles.
3. **a.** A parasite is an organism which lives in or on another organism and damages it; this organism is called the host.
   b.

Reasons to classify viruses as living organisms	Reasons not to classify viruses as living organisms
They reproduce in large numbers.	Viruses cannot carry out any of the seven characteristics of living organisms on their own. They can only reproduce once they are inside a living organism such as an animal or plant.
They need food, energy, etc. and they get them from the organisms they invade so they are perfect parasites.	Some viruses can survive for years stored in airtight containers and still cause disease years later.

1.10 Moving classification forwards

1. **a.** Scientists collect organisms and then make careful observations of characteristics such as number of limbs, how it reproduces, number and colour of flowers, etc. They compare this with known organisms to see if it is a new species.
   b. Any sensible two points, e.g. it is cheap, it is widely available anywhere in the world; it does not depend on expensive technology or a supply of electricity, etc.
2. **a.** It avoids misclassifying species as new when they are not, or old when they are new. It is rapid and can be done in the field/ avoids taking specimens out of the wild.
   b. Any two sensible points, e.g. it is relatively expensive, it is not widely available everywhere in the world; it requires relatively expensive technology or a supply of electricity, people need special training to use it, etc.
3. You should use knowledge from the student book and from the text here to write a short essay – your work should have a good structure, clear ideas with explanations, and interest.

2.1 The building blocks of life

1. The missing words are: cells; bacteria; billions; seven; respiration/excretion/growth (order of these three doesn't matter); 20–30; microscope.
2.

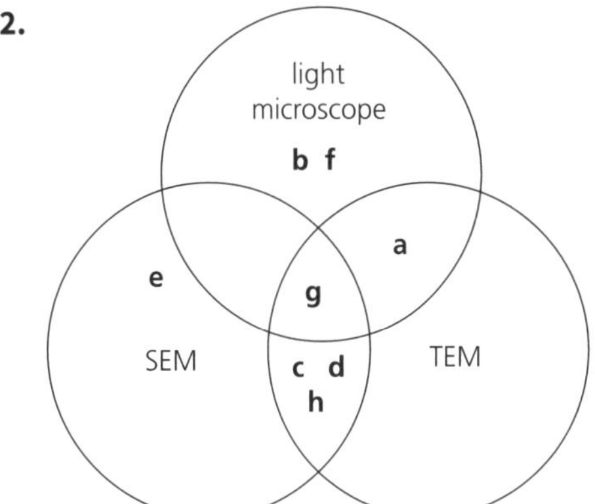

Extension

a. e.g. magnify up to 1 million times; enable us to see a lot of detail inside cells.

b. e.g. very large, very expensive, can't be used in schools or in the field.

2.2 The cell story

1. The first microscopes were made in the 17th century CE.
2. The first material seen under a microscope by Robert Hooke was a thin slice of cork.
3. Anton van Leeuwenhoek was excited when he looked down his microscope at a drop of pond water

because he saw living microscopic animals moving around. He was the first person to see living cells.

4. There were two main reasons: for many years microscopes were not very good so very few people actually saw any cells; and for many years communication between countries was not very good so scientists in one place did not know what was happening elsewhere.

5. The cell is the basic unit of life/basic unit of structure of all organisms. All organisms consist of one or more cells.

Extension

You should write fluently. Show you understand that Leeuwenhoek made lenses to see his cloth clearly. Note his pleasure/pride in using the lenses to make a microscope. You should imagine and describe his excitement at seeing tiny living organisms swimming about.

2.3 Animal and plant cells

1. A nucleus, B cell membrane, C cytoplasm, D mitochondria.

2. A cell wall, B cell membrane, C vacuole, D nucleus, E cytoplasm, F mitochondria, G chloroplasts

3. A3; B5; C7; D1; E4; F2; G6

Extension

Muscle cells – they do a lot of work moving the body around so they need a lot of energy and need a lot of mitochondria. Fat cells need less energy so have fewer mitochondria.

2.4 Using a microscope

1. Eyepiece A, objective lens B; stage C, slide D, coarse focus E, fine focus F, light source G.

2. The correct order is:
   - Move the stage of the microscope to its lowest position.
   - Choose the objective lens with the lowest magnification and move it into position.
   - Put the object you want to look at on the stage. Keep it in place using the clips.
   - Look through the eyepiece lens and turn the coarse focus knob slowly until the object comes into focus.
   - Once the object is in focus , turn the fine focus knob slightly to see if it makes the image even sharper.
   - To see your specimen in more detail, repeat these steps using an objective lens with a higher magnification.
   - Draw your observations using a pencil. Record the magnification you are using.

3. total magnification = $5 \times 4 = \times 20$

Extension

total magnification = eyepiece lens magnification × objective lens magnification. $\times 100 = \times 4$ objective lens magnification. Rearrange to calculate objective lens magnification = $100/4 = \times 25$.

2.5 Specialised animal cells

1. The missing words are: cells; respiration; multicellular; specialised; structure; function.

2. A3; B1; C2; D5; E4

3. Filled with haemoglobin: red substance/molecule that carries oxygen.

   Small and flexible: can pass through tiny blood vessels to carry oxygen to the cells.

   No nucleus: makes more space for haemoglobin to carry oxygen.

   Biconcave: gives a large surface area to pick up oxygen.

Extension

Main adaptations of ciliated cells: cilia – tiny hair like structures on the edge of the cell – and many mitochondria. Cilia beat to move things about in the body. This movement needs energy, so ciliated cells have lots of mitochondria to provide the energy from respiration needed for the cilia to beat.

2.6 Specialised plant cells

1. a. Photosynthesis.
   b. (Adaptations can be given in any order) Adaptation/explanation 1: Cells are located near top of the leaf/so they get as much sunlight as possible. Adaptation/explanation 2: cells are brick-shaped/so they can be packed together as tightly as possible. Adaptation/explanation 3: they contain large numbers of chloroplasts/to capture light for photosynthesis and make food.
2. a. Root hair cell.
   b. Take in water and mineral salts from the soil.
   c. Your labels should be the same as Fig 2.6C in your Student Book. Your notes on the diagram should include: long microscopic hair to give a big surface area to take in water and mineral salts; found on outside of the root – to grow into the soil and reach the soil water; large sap-filled vacuole which helps move water from the soil into the root.

Extension

a. Chloroplasts in palisade cells.
b. Palisade cells found in top layers of leaf where they get a lot of sun. Chloroplasts capture light to use in photosynthesis and make food for the plant. Root hair cells grow under the ground where there is no light – they don't need chloroplasts as there is no light to capture and so they cannot photosynthesise.

2.7 Modelling cells

1.

Feature	Animal cells	Plant cells
cell membrane	✓	✓
cell wall	✗	✓
nucleus	✓	✓
large sap vacuole	✗	✓
chloroplasts	✗	✓
cytoplasm	✓	✓
mitochondria	✓	✓
carries out respiration	✓	✓
carries out photosynthesis	✗	✓

2. The missing words are: physical; explain; big/small; small/big; misconceptions; remember.
3. The same: any sensible points, e.g. have a nucleus, have cytoplasm, have cell membrane, have mitochondria.

   Different: Must include – plant cells are bigger than animal cells; any other sensible points, e.g. plant cells are more regular in shape than animal cells; plant cells have a cell wall/chloroplasts/large vacuole and animal cells do not.

2.8 Tissues and organs in animals

1. A2; B4; C1; D5; E3
2. Brain: controls the body; lungs: take in oxygen and remove carbon dioxide; heart: pumps blood around the body; stomach: digests food; liver: removes toxins (poisons) from the body; intestine: absorbs nutrients from food; kidney: filters the blood and produces urine; bladder: stores urine.

Extension

Depends on which organ system you chose. Your answer should show awareness of the organs making up the organ system and some of the key tissues, e.g. muscle tissue to move food through the digestive system.

2.9 Tissues and organs in plants

1. The missing words are: multicellular; organisation; cells; tissues; organs; whole organism; two; shoot; photosynthetic; root.
2. a. A leaf; B stem; C roots
   b. Stem holds the plant upright; root anchors the plant in the ground and takes up water and mineral salts from the soil; leaf absorbs sunlight for making food by photosynthesis.
3. a. flower. b. reproduction.
   c. Because it is not always present/plants only flower at certain times of the year.
4. a. xylem. b. phloem. c. palisade tissue.

3.1 Microorganisms

1. The missing words are: single; bacteria/viruses; viruses/bacteria; fungi; microscope; culture.
2. **a.** fungus **b.** bacteria **c.** viruses **d.** yeast/s.
3. **a.** C
   b. F, plasmids.
   c. **i.** Any two from: A cell membrane, B cytoplasm, C loop of genetic material, E cell wall.
   ii. Corresponding two from: cell membrane – controls the movement of substances in and out of the bacterial cell; cytoplasm – site of most of the reactions that take place in the cell; loop of genetic material – carries information about making new bacteria; cell wall – gives the cell strength.

Extension

Your table should look something like this:

Feature	Bacteria	Fungi	Viruses
Size	0.2–2.0 μm	4.0 μm	0.01–0.1 μm
Genetic material	loop of genetic material + plasmids	nucleus	strand of genetic material
Cell wall	different to plant and fungal cell walls	present – made of chitin	no cell wall – protein coat
Cell contents	cytoplasm, cell membrane, loop of genetic material	cytoplasm, cell membrane, nucleus containing genetic material	genetic material

3.2 Microorganisms are our friends

1. The missing words are: dough, flour, yeast, respiration, carbon dioxide, soft.
2. **a.** yoghurt, cheese.
   b. Any antibiotic, e.g. penicillin/streptomycin. Antibiotics stop bacteria growing or kill them and so they cure infectious diseases caused by bacteria.
3. **a.** The milk with extra bacteria added had a lower pH after 4 hours.
   b. Bacteria use the sugars in milk for respiration and produce lactic acid as a waste product. Lactic acid lowers the pH of the milk mixture and gives the yoghurt a sharp, acidic taste.

Extension

a. Graph as shown below.

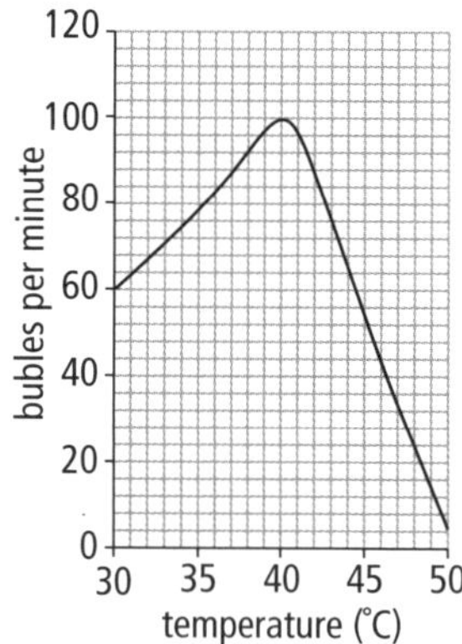

b. The number of bubbles per minute increases as the temperature increases until it reaches 40 °C, because yeast respire faster at higher temperatures. Above 40 °C, there are fewer bubbles per minute because the higher temperatures destroy the yeast.

3.3 Microorganisms and disease

1. The correct order is:

   2 Pasteur decides to grow anthrax germs, weaken them, and inject them into animals to protect them, against the disease and prove his germ theory.

   1 Pasteur has a hypothesis that infectious diseases like anthrax are caused by germs passed from one person or animal to another.

   7 Three days after the animals were infected with anthrax, there was a public inspection. All the vaccinated animals were alive and healthy. All the unvaccinated animals were dead or dying of anthrax.

   3 Hippolyte Rossignol does not believe in germ theory and challenges Pasteur to prove his theory with a public test.

   5 A few weeks later all the animals are infected with live anthrax from infected animals.

   6 Pasteur predicts that the vaccinated animals will survive and the unvaccinated animals will develop anthrax and die.

   8 Pasteur had clearly demonstrated both the germ theory of disease and the value of vaccination.

   4 Pasteur gives a group of healthy sheep, cows and goats his vaccine. He selects a similar group of animals that do not get the vaccine.

Extension

Says: When microorganisms land in liquids full of nutrients they grow and reproduce, and make the liquid cloudy – suggest an explanation.

Thinks: if I keep microorganisms out of a nutrient solution it will not go cloudy because it will not go bad – make a prediction.

Places nutrient solutions in flasks with S-shaped necks, and boils them to destroy any microorganisms present – test the explanation.

Notices that nutrient solutions do not go cloudy in flasks with S-shaped necks – review the evidence.

Breaks the neck of one of the flasks and observes that the nutrient solution begins to ferment and turn cloudy once microorganisms can get in – collect extra evidence.

3.4 Using science to prevent disease

1

How disease-causing microorganisms spread from person to person
Droplet infection: Tiny droplets containing disease-causing microorganisms forced out of your nose and mouth when you breathe, talk, cough, or sneeze. Other people breathe in the droplets and take the microorganisms into their body.
Contact: Some microorganisms passed from one person to another by direct contact with skin or with other objects people have coughed/sneezed over, etc. You transfer the microorganisms to your mouth, eyes, etc. by touch.
Contaminated food and drink: Water contaminated by human waste or raw or undercooked food may be contaminated with disease-causing microorganisms. The microorganisms are taken directly into the body.
Break in the skin: Some microorganisms get straight into the blood stream through cuts or breaks in the skin, including animal bites and needle scratches.

2. **a.** Wear a face mask to protect your mouth and nose and prevent you breathing out/coughing out/sneezing microorganisms into the air for other people to take in; avoid crowded places so there are fewer microorganisms in the air; cough or sneeze into a tissue or your elbow to prevent a microorganism spray; meet outdoors where there is less build up of microorganisms in the air; any other sensible points.

   b. Wash hands often/use antiseptic gel to prevent transfer of microorganisms from surfaces to mouth; don't shake hands; maintain distance from people so you don't touch them to transfer microorganisms; any other sensible points.

Extension

Any three from: improve sewage treatment so that drinking water is never contaminated with faeces carrying typhoid bacteria; make sure everyone can get clean drinking water so people are not forced to drink water that may be contaminated with typhoid bacteria; encourage people to wash their hands after using the toilet to make sure they don't transfer typhoid bacteria to surfaces, food, etc.; encourage people who sell or prepare food not to work when they feel ill to minimise risk of them passing on typhoid bacteria on their hands; any other sensible point with scientific explanation.

3.5 The decomposers

1. The missing words are: water; mineral salts; herbivores; dead; decomposers; bacteria/fungi; fungi/bacteria; nutrients; plants.
2. **a.** Decomposers are organisms that break down animal droppings, plant fruits and leaves that fall to the ground, and dead animals and plants. They use some of the nutrients to grow and reproduce and return some to the soil.
   b. Compost is a brown, nutrient-rich material made when kitchen and garden waste is broken down by decomposers.
   c. Plants take mineral salts from the ground and use them to build new plant tissue. Plants are eaten by herbivores. Eventually all the mineral resources of the soil would be used up. Compost returns many of the minerals to the soil and keeps it fertile.

Extension

a. The human population is growing all the time and more people produce more human waste.
b. In pit latrines or in modern sewage plants, we use decomposers to break down and digest human waste, making it safe to use as fertiliser and making the water clean enough to return to rivers.
c. Human sewage is damaging to the environment and it may carry diseases.

3.6 Investigating rotting rates

1. **a.** An anomalous result is one that does not fit in with the pattern of the rest of your results.
   b. Three main stages are: evaluating the quality of the data, e.g. anomalous results, the spread and the range, are there random or systematic errors? Evaluating the methods used, e.g. measuring methods used. Suggesting improvements to your investigation to improve the quality of the data collected so you can be more confident in your conclusion.
   c. Random errors are the result of unexpected changes, etc. – they cause anomalous results. Systematic errors are the result of a damaged piece of equipment which shows the same error in every reading.
2. **a.** Tomatoes decompose faster at room temperature than in the fridge. Tomatoes decompose faster if they are exposed to the air whether in the fridge or at room temperature.
   b. The method is not strong – different masses of tomatoes, no accurate measurements of temperatures, cover of mould is a subjective measure of decomposition, etc.
   c. Measure the mass of tomatoes as see how it changes, measure the temperature of the room and fridge – any other sensible points.

Extension

The best option is a bar graph with bars for each day clustered together. Use different colours or patterns for different treatments; suitable scales; labelled axes, etc.

3.7 Food chains, food webs and decomposers

1. A2; B3; C1; D5; E4; F7; G6
2. **a.** maize
   b. mice
   c. snakes
   d. This is the direction in which biomass is transferred through the food chain. The plant has the most biomass because it makes food by photosynthesis. Biomass is lost at each stage of the chain.

e. Decomposers arrow goes from mongooses back to maize with label/explanation that decomposers break down biomass and return mineral salts to the soil to be taken up and used by the plants.

3. a. A food web is a model of the feeding relationships between organisms in an ecosystem. It links a number of food chains together.
   b. Most animals eat more than one kind of organism so a food web is a more realistic representation of the feeding relationships.
   c. **Herbivores**: two from ant, mouse, lizard; **carnivores**: two from spider, scorpion, large lizard, snake, fennec fox/fox.
   d. Producers/plants are missing – they make biomass by photosynthesis so they are the starting point for the whole food web; decomposers – they break down and digest the droppings and dead remains of animals and plants in the food web, returning mineral salts to the soil to be used and recycled by plants.

4.1 The physical environment

1. The missing words are: abiotic; rocks; temperature; water; fertile; cold; water.
2. a. Any two, e.g. transport of substances around body, in photosynthesis, in sweating, to remove waste, etc.
   b. Highest percentage water = lungs; lowest percentage water = bones.
   c. 82–31; 51% +/−1

Extension

a. All of the reactions of life take place in solution in water.

b. This stops the water under the ice from freezing, so the organisms living below the ice survive until it gets warmer and the ice melts.

c. The temperature of oceans and big lakes is very stable so organisms do not have to cope with big temperature changes.

4.2 The water cycle

1. A melts; B boils/evaporates; C freezes; D condenses.
2. Your answers should be in complete sentences similar to the following:
   a. **Evaporation**: Energy from the sun heats up the water in puddles, streams, lakes, rivers, and oceans. Some of the water evaporates. It turns from a liquid to a gas called water vapour.
   b. Water vapour rises up into the sky. The higher it gets, the cooler it is. When the water vapour cools down, it turns back into liquid water. This is **condensation**. The small drops of liquid water form clouds in the sky. These clouds can travel thousands of miles up in the atmosphere.
   c. **Precipitation**: The water droplets in the clouds get bigger and heavier as more water condenses. When they get big enough, they fall back down to Earth as precipitation, which may be rain, snow, hail, or sleet depending on the temperature.
   d. **Collection**: Precipitation collects in bodies of water such as oceans, lakes and rivers, or it soaks through the ground and collects as groundwater, or it runs off the surface of the soil into bigger bodies of water.

Extension

Based on the water cycle. Molecules of water cycle through the environment over hundreds or thousands of years. As part of the water cycle, they spend time in the bodies of plants and animals. So a molecule of water we drink today may have been drunk by a dinosaur millions of years ago.

4.3 Global warming and the water cycle

1. T; F; F; T; T; F; T
2. When fossil fuels are burned in car engines and electricity generation it produces carbon dioxide which is warming the Earth. This is having a global impact on sea levels as (1) the polar land ice and glaciers are melting into the sea so the levels rise and (2) as the sea gets warmer it expands and so sea levels rise.

3. a. The increase in the surface temperature of the Earth means more water evaporates from the surface into the air. Warm air also holds more water vapour. As this water vapour rises and cools, it forms bigger clouds because there is more of it. This makes bigger rainstorms and so there is more risk of flooding.
   b. In some areas the water evaporates from the soil as a result of global warming but there are also strong winds so the clouds that form are blown away. No rain falls, causing droughts.

5.1 Diffusion in biology

1. The missing words are: particles; diffusion; gases/liquids; liquids/gases; random; net; particles; high; lower; down; concentration gradient.
2. Any three, e.g. sense of smell relies on diffusion, pollinators finding flowers, gas exchange in the lungs, food moving from digestive system to the blood, sharks detecting prey, etc.
3. a. Dark particles should have spread out a bit further and be more mixed with light particles but there will still be an area which has only light particles.
   b. The whole volume of liquid has evenly spaced dark and light dots.

Extension

Diffusion is the net movement of particles down a concentration gradient, from a high concentration to a lower concentration. At the beginning there is a very high concentration of potassium manganate VII particles around the crystal and a high concentration of water particles everywhere else. Random movements of the particles mean the potassium manganate VII particles move down a concentration gradient into the areas of water only particles, and the water particles show a net movement down a concentration gradient toward the potassium manganate VII crystal. After 24 hours, the random movements will result in an even mixing of the two types of particles. The random movements will continue but there will be no more net movements.

5.2 Aerobic respiration in animals and plants

1. The missing words are: : energy; aerobic; energy; glucose; controlled; oxygen; carbon dioxide; water.
2. glucose + oxygen → carbon dioxide and water
3. a. A is an animal cell and B is a plant cell.
   b. Aerobic respiration takes place in the mitochondria. Check you have correctly labelled them in both animal and plant cells by comparison with your Student Book.

Extension

Cells such as muscle cells in animals and the cells that produce seeds and fruits in plants need a lot of energy to carry out their functions. This energy comes from the controlled breakdown of glucose using oxygen in aerobic respiration. This takes place in the mitochondria. Cells that need a lot of energy have lots of mitochondria to supply the energy. Cells like storage cells use little energy so they need few mitochondria.

5.3 Anaerobic respiration

1. a. Glucose + oxygen → carbon dioxide + water
   b. Glucose → lactic acid
2. **a.** T **b.** F **c.** T **d.** F **e.** F
   b. **Anaerobic** respiration provides only short bursts of energy or Aerobic respiration **provides a continuous supply of energy**.
   d. Anaerobic respiration releases **a smaller** percentage of the total energy in glucose **than** aerobic respiration.

e. **Anaerobic** respiration is the main type used in **sprinting races**.

3. a. Labels should be added to the diagram as shown below.

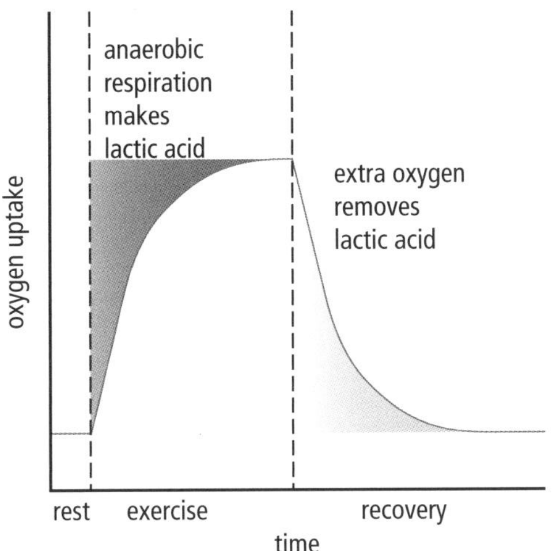

b. Anaerobic respiration can't be used all of the time because lactic acid is toxic.

c. Anaerobic respiration can produce a sudden burst of energy because cells can use a lot of glucose molecules at once so, even though the anaerobic breakdown of each molecule only releases a small amount of energy, a lot of molecules are all used to give the energy needed.

5.4 Investigating respiration

1. a. They are hazard warning signs.

b. ! Harmful substance.

c. Limewater is a clear, colourless liquid (a dilute solution of calcium hydroxide). It turns milky white in the presence of carbon dioxide. Carbon dioxide is a waste product of aerobic respiration so it is used to show living organisms are respiring and producing carbon dioxide.

2. a. That people who have been exercising will produce more carbon dioxide than people at rest.

b.

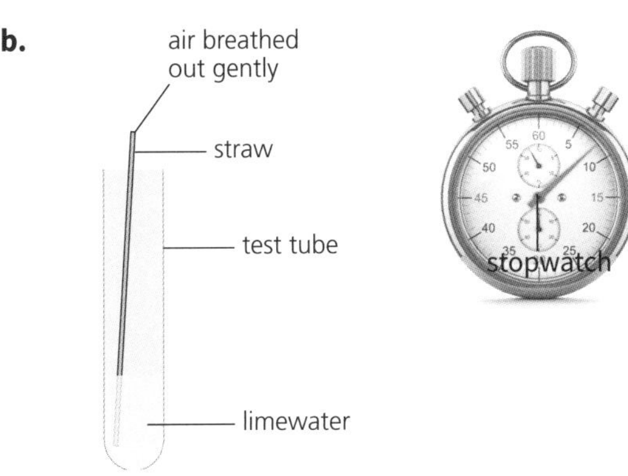

c. Controlled variables – resting or exercise, time of exercise, amount of limewater in test tube. Variables which are hard to control: differences between individuals, levels of fitness, etc.

d. Expected results: The limewater turns cloudy faster when people who have been exercising breathe through it.

Explanation: When people exercise, they need more energy for their muscles to work. Their cells carry out more respiration to provide this energy and produce more carbon dioxide. So the air they breathe out through the limewater after exercise has a higher concentration of carbon dioxide than when they breathe out through limewater at rest. The limewater turns cloudy faster with the higher concentration of carbon dioxide.

Extension

a. Control – sealed tube containing limewater and an empty muslin bag.

b. Control – limewater stays the same as nothing is respiring so no carbon dioxide is being produced. Germinating seeds and worms – the limewater gradually turns cloudy, reacting with the carbon dioxide produced by respiration. The limewater would go cloudy faster in the tube with the worms as they are animals and move about and use more energy than germinating seeds, so they will produce more carbon dioxide, turning the limewater cloudy.

5.5 The lungs and gas exchange

1. Clockwise from the top right the labels are: trachea, bronchus, small branch of bronchi, alveoli, diaphragm, lung, rib.

2. The missing words are: oxygen; aerobic; respiratory system; air; lungs; gas exchange; oxygen; carbon dioxide.

3. A5; B4; C1; D6; E3; F2

Extension

Specialised cells lining the bronchi make mucus that traps dust and dirt from the air that might irritate the lungs and harmful microorganisms that might cause lung infections. They are ciliated and the cilia beat to move the mucus, dirt and microorganisms away from the lungs.

5.6 Breathing

1. The missing words are: blood; alveoli; energy; glucose; cells; exercise; exchange; oxygen.
2. Oxygen in air breathed out = 16, carbon dioxide = 4. Your answer should total close to 100% and show there is still plenty of oxygen in the air breathed out.
3. 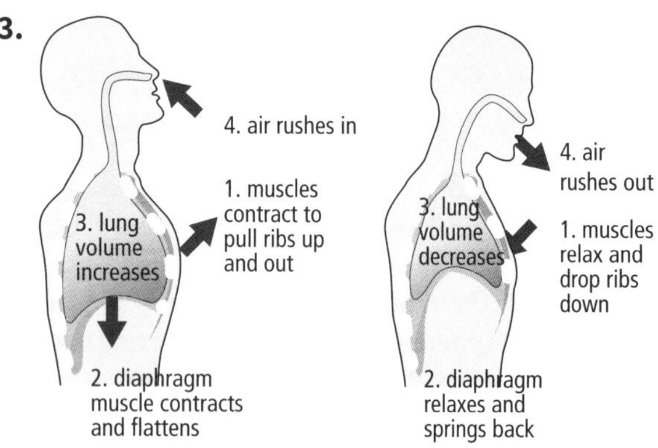

Extension

Lungs are not single sacs like balloons – lungs are made up of millions of tiny air sacs called alveoli so a sponge is a better model. A balloon could model a single alveolus.

Air is not blown into the lungs – changes in volume and pressure in the lungs mean air is forced into the lungs by external pressure.

Any other sensible points.

5.7 The effect of exercise on the breathing rate

1. **a.** Predict the breathing rate of students in Group A will be slower than students in Group B.
   b. Expect the breathing rate of all the students to increase. Expect the breathing rate of Group A students to increase less than the breathing rate of Group B students.
   c. When people exercise they need more oxygen for aerobic respiration in their muscles, so everyone's breathing rate goes up. As people get fitter, their lungs get bigger and more efficient so they cope better with the demand for extra oxygen and their breathing rate doesn't go up as much.
2. **a.** 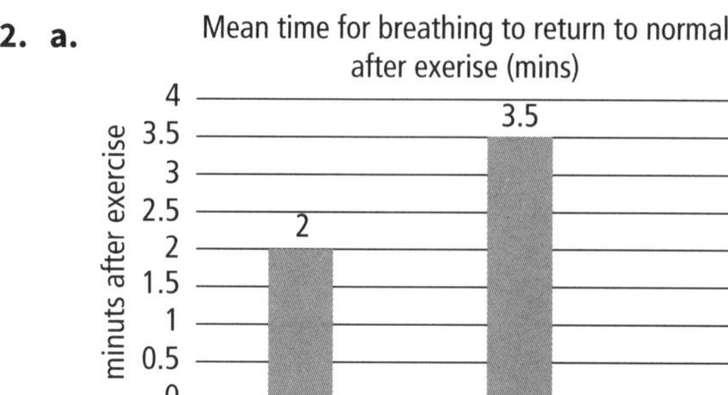

   b. It takes Group B students on average 1.5 minutes longer for their breathing to return to normal than students in Group A.
   c. **i.** The breathing rate of students who do regular exercise recover more quickly from exercise than students who do less.
   ii. Any sensible suggestions, e.g. don't know how many students are in each group; don't know how the investigation is carried out in terms of timing, etc.; don't know if there are equal numbers of students in each group; don't know how many times each investigation was repeated.

Extension

Your answer should show that you recognise that the results given are means so they are the individual results of many students analysed, and that there will be a lot of variation between individuals and therefore between results in each group. Some students within each group will be fitter/have bigger or smaller lung capacities, etc.

5.8 The structure of the alveoli

1. gas exchange; alveoli; adaptations; diffuses; air; blood; dioxide; capillaries; alveoli.
2. 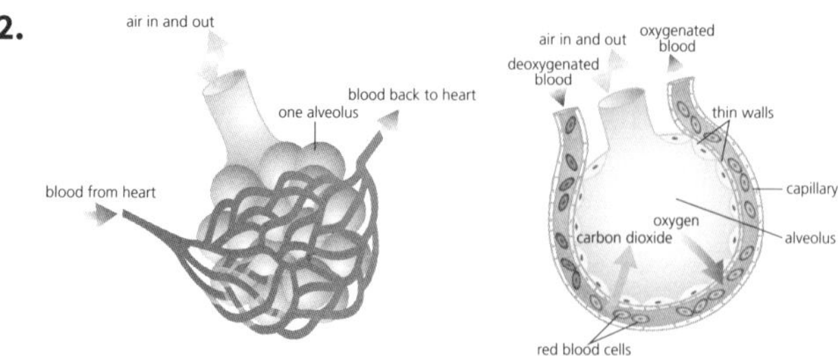

3. a. Many tiny air sacs give a very big surface area over which gas exchange takes place.
   b. Very short distances for the oxygen and carbon dioxide to diffuse.

Extension

a. The steep concentration gradient between oxygen in the air and the oxygen in the blood means oxygen diffuses quickly down the concentration gradient from the air in the alveoli to the blood in the capillary. The steep concentration gradient between the carbon dioxide in the blood and in the air in the alveoli means the waste gas diffuses out of the air into the blood as fast as possible.

b. The rich blood supply helps maintain the steep concentration gradients needed for the rapid diffusion of oxygen into the blood and carbon dioxide out of it.

5.9 Asthma

1. a. The muscles around the airways to the lungs contract making the tubes narrower; the linings of the bronchi swell and make more mucus during an asthma attack.
   b. Both of these changes make the airways narrow and make it hard to move air into and out of the lungs. This means that the person affected does not get enough oxygen so they feel breathless.
2. a. Any two from the following: Lotanna breathes out a larger volume (or the reverse for Maryam); Lotanna breathes out faster (or the reverse for Maryam); Lotanna empties her lungs faster (or the reverse for Maryam).
   b. Maryam has asthma.
3. Scientists had to discover what happens in an asthma attack and what causes asthma before they could begin to develop medicines. Once they had developed medicines that stopped the muscles contracting and the tubes swelling, they needed to get the medicines into the respiratory system where it is needed. So they had to apply scientific principles from engineering to develop an inhaler to deliver the drug to where it is needed.

5.10 Transport in the blood

1. a. Platelets, red blood cells, plasma, white blood cells.
   b. Red blood cells are the most common – they are very small and need huge numbers to carry oxygen around the body.
2. **a.** Red blood cells. **b.** White blood cells. **c.** Red blood cells. **d.** White blood cells. **e.** Plasma. **f.** Plasma. **g.** Platelets. **h.** Red blood cell.
3.

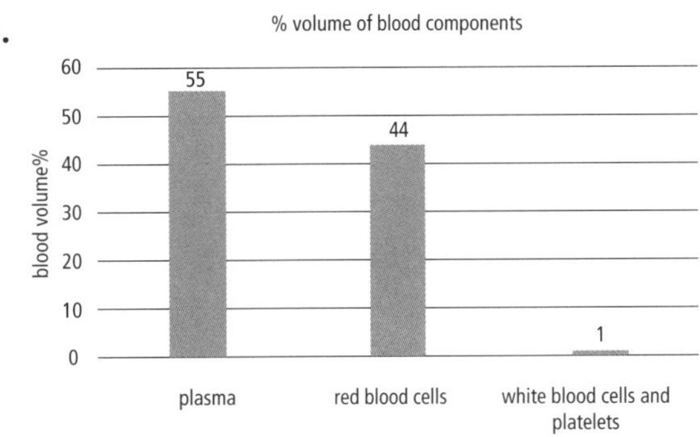

Extension

a. Red blood cells carry oxygen needed for cellular respiration to provide energy to carry out the processers of life. Lack of red blood cells means lack of oxygen so less energy which results in feeling tired.

b. Red blood cells contain haemoglobin which carries oxygen around the body. A patient with anaemia has fewer red blood cells than normal and so they will have less haemoglobin.

c. The body needs iron to make haemoglobin for the red blood cells. A lack of iron in the diet means less haemoglobin can be made so fewer red blood cells will be made, causing anaemia.

5.11 Coping with extremes

1. a. Large rib cages, many more small airways and alveoli in the lungs; more capillaries in the lungs; many extra red blood cells.
   b. Large rib cages – take more air in with each breath; many more small airways and alveoli in the lungs – bigger surface area for gas exchange; more capillaries in the lungs – more opportunity for gas exchange with the blood, maintains steeper concentration gradients; many extra red blood cells – take more oxygen from the air and haemoglobin in red blood cells attaches to oxygen. Any other sensible points.

2. a. Bradycardia/slowing of the heart rate.
   b. Reduces the amount of oxygen delivered to and used by the body tissues so the oxygen lasts longer.
   c. Any two from: Apnea/stops breathing: prevents drowning by bringing water into the lungs; capillary shutdown: saves blood and oxygen for the organs where it is needed; big myoglobin stores in the muscles: oxygen stores in the muscle means they can respire aerobically for most of the dive without taking it from the blood.

3.

Glycogen	Fat
a short-term carbohydrate energy store stored in our muscles and liver used for quick, instant energy	a long-term energy store stored in special cells under the skin and around the body organs broken down for energy over time if not enough food is eaten

6.1 The food we eat

1. A3; B4; C5; D2; E1
2. **a.** Carbohydrates. **b.** Fats. **c.** Carbohydrates and fats. **d.** Proteins and fats.
3. Fats are needed: for insulation, for making cell membranes, and as a source of energy.
4. digestive system; large, insoluble; small, soluble; diffusion; cells.

Extension

Their diet lacks protein so their growth may be slowed (stunted)/ if they get any injuries they will not heal easily.

6.2 Carbohydrates, fats, and energy

1.

Nutrient	Small molecules joined to make them
starches	sugars
proteins	amino acids
fats	fatty acids and glycerol

2. **a.** F; **b.** T; **c.** F; **d.** F; **e.** T
   b. Glucose is an example of a **sugar/carbohydrate**.
   c. Fats and oils contain **more** energy per gram than carbohydrates.
   d. Each fat or oil has a different set of **fatty acids** attached to a glycerol molecule.

Extension

All of these foods are high in fats and many are high in sugars. This combination often tastes very good, so we want to eat a lot of it BUT fats and sugars are very high in energy. It is easy to eat more of these foods than we need and our bodies store the extra energy as fat in our fat cells.

6.3 Measuring the energy in food – managing variables

1. **a.** Changed. **b.** Controlled. **c.** Controlled. **d.** Controlled. **e.** Measured. **f.** Measured. **g.** Calculated.

2.

Measuring instrument	Variable measured	Units used
thermometer	temperature	°C
measuring cylinder	volume	cm^3
electronic balance	mass	g

3. Any two from: a large amount of food would make the water boil (so she could not calculate an accurate temperature rise); some of the heat would escape around the sides of the test tube; pieces of burning food could fall off and be a hazard.

Extension

a. Bread produces a lower temperature rise per gram because it contains less fat than chicken or cheese.

Temperature rise (°C)	Temperature rise per gram (°C per gram)
30	20
46	46
66	33

b. Any two from: use the same mass of each type of food; make sure the initial water temperature is the same for each food type; repeat each experiment several times to find the mean temperature rise per gram; any other sensible suggestion.

6.4 A balanced diet

1. The missing words are: nutrient, proportions, proteins/lipids, lipids/proteins, minerals, rice, energy, fibre/water, water/fibre.
2. Chocolate is high in sugar which releases energy quickly but it also contains a lot of fats so it is very high in energy – eating a lot of chocolate to supply the carbohydrates you need would lead to weight gain because of all the energy in the fat. It does not contain many other important nutrients, e.g. protein. Any other sensible point.,
3. **a.** A model which shows the proportions of the different types of food that should be eaten in a balanced diet.

   b. **Bottom layer**: carbohydrates, e.g. bread, rice, pasta, etc. **second layer**: fruit and vegetables, e.g. banana, lemons, mangos, yams, etc. **third layer**: protein, e.g. meat (or specific meats), eggs, cheese, etc. **top layer**: fats and oils, e.g. butter, cream, different oils, etc.

Extension

Any three factors from: age, level of activity, climate where you are living, if you are pregnant, if you are ill, any other sensible suggestion.

For example, age: children need plenty of energy and a lot of protein, calcium, and vitamin D as they are growing new bone and muscle. Older people need less energy from food as they are less active but need plenty of minerals, vitamins, and protein to keep repairing their body; young active people need plenty of carbohydrates to supply their energy needs. People in cold countries need more energy than those in hot countries to stay warm. Any other sensible points.

6.5 Diet, growth, and development

1. The missing words are: ill, explain, scurvy, deficiency, lack, vitamins, nutrients.
2. Anaemia – iron; night blindness – vitamin A; scurvy – vitamin C; rickets – vitamin D AND calcium.
3. **a.** Hassina may have scurvy and anaemia.

   b. Scurvy would cause bleeding gums, swollen legs, and a lack of energy. Anaemia would cause tiredness, weakness, and a lack of energy.

   c. Scurvy: eats lots of citrus fruits and tomatoes which are rich in vitamin C.
   Anaemia: eat food rich in iron, e.g. red meat, egg yolks, apricots.

Extension

a.

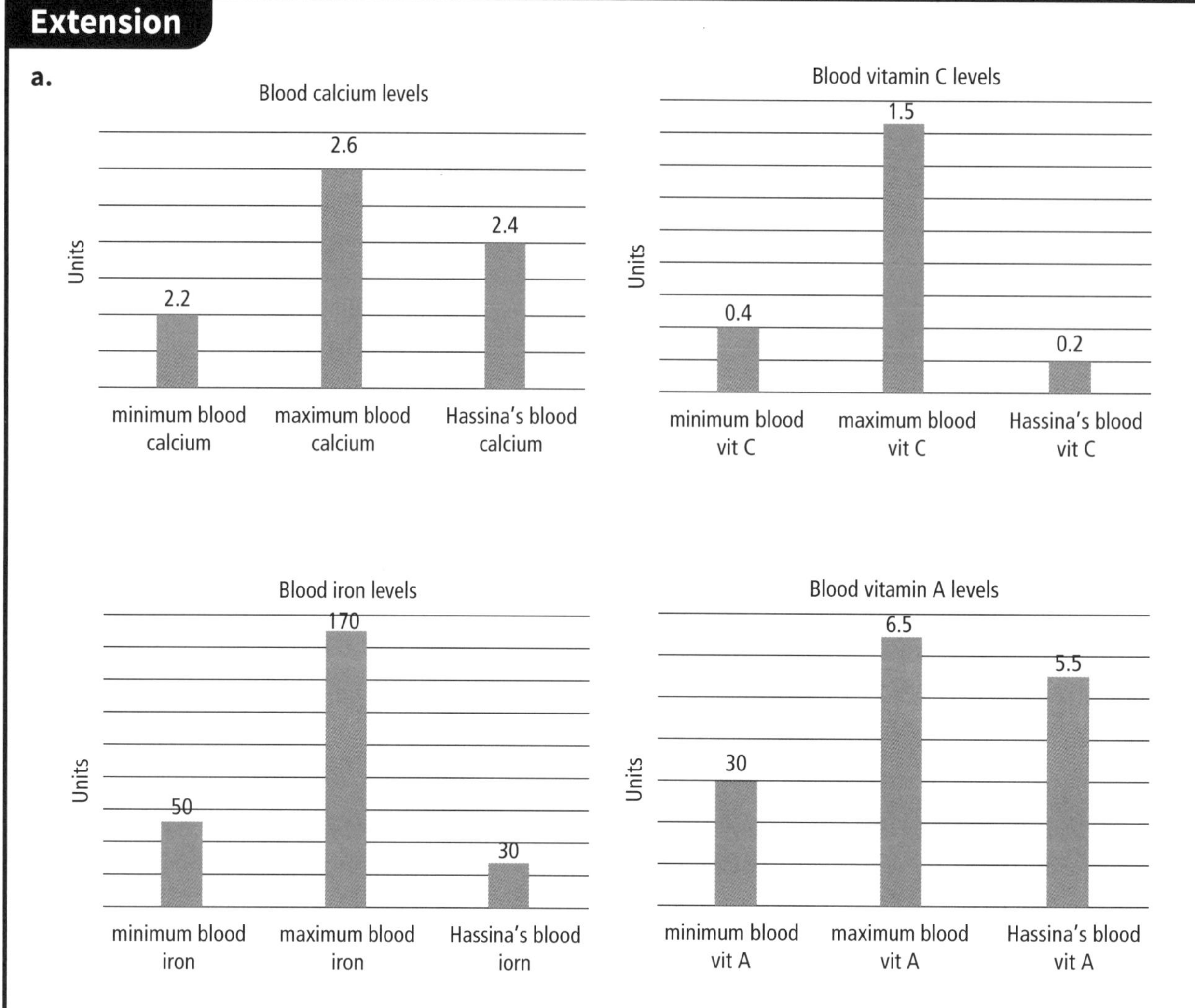

b. The scales/amounts of the different substances are very different. If they were all on the same graph, you wouldn't be able to see the differences clearly.

c. Displaying the data on these bar charts makes it very easy to see whether Hassina's blood levels of the different minerals and vitamins come within the normal range.

6.6 Starvation, obesity, and health

1. Underline: too thin; waste away; little; deficiency; fail to grow; do not; Starving; ill; is not; million.
2. The missing words are: body, obese, energy, fat, diabetes, blood, heart.
3. A3; B1/4; C2; D4/1
4. Diabetes: problems controlling the blood sugar levels which cause tiredness, circulation problems, blindness, and eventually death.

   Heart disease: fatty material builds up in the blood vessels which may cause a heart attack.

   Arthritis: weight causes strain on the skeleton leading to wear and pain in the joints.

6.7 Smoking and health

1. **a.** Labels for the diagram should carry this information:

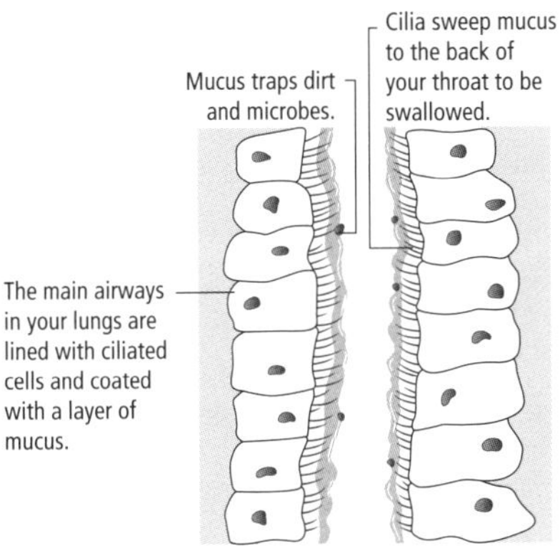

   b. Any six from the following: cilia are paralysed (stop moving) so do not move mucus away from the lungs; dirt, microbes, and chemicals from cigarette smoke build up in the mucus; lung infections increase; mucus is coughed up; the blood vessels become narrower; breathing becomes difficult; the alveoli walls break down; the surface area available for gas exchange is reduced/gas exchange is less efficient/less oxygen taken up in the lungs.

   c. Any two from: lung cancer, heart disease, COPD, throat and mouth cancers.

2. The missing words are: harmful, cilia, cancer, reduces, raises, narrower, heart, addictive.

Extension

Mohamed is the smoker. His race time is slower because the alveoli of his lungs are coated with tar which makes gas exchange more difficult. He gets less oxygen into his blood and so his muscles get less oxygen. They cannot respire as effectively and so he gets less energy. His muscles don't work as well and tire more easily. This is made worse because up to 10% of his blood is carrying carbon monoxide instead of oxygen, so his blood carries even less oxygen to his muscles.

6.8 Building the evidence

1. The missing words are: tobacco; smoking; lung cancer; observed; lifestyle; smokers; results; peer-reviewed; evidence; causal link; tobacco; risk.

2. **a.** Non-smoker has 10/100 000 risk of developing lung cancer.

   Smoking 25+ cigarettes a day gives 350/100 000 risk of developing lung cancer.

   350/10 = 35 so someone who smokes 25+ cigarettes a day has a 35× higher risk of developing lung cancer than a non-smoker.

   b. Tells us that lung cancer does occur in non-smokers so it is not always caused by smoking but that smoking greatly increases the risk.

   c. A peer-reviewed journal is a journal where the paper or articles are looked at by other scientists in the field before they are published. This is important as it makes sure that the findings are checked and inaccurate science is not published.

   d. Any sensible point, e.g. the number of people in the survey (the more people, the more reliable the data), who carried out the survey (recognised scientists mean the data is more likely to be reliable), where the survey was done (good facilities and accurate recording important), how long the people in the survey had been smoking (if some had only just started and others been smoking for years, the data is less relevant).

6.9 The human skeleton

1. **a, b.** Support, e.g. vertebral column holds body upright; protection, e.g. skull protects brain / ribs protect heart and lungs/ backbone/vertebrae protect the spinal cord; movement joints where bones meet allow movement, e.g. hip joint, knee joint; making blood cells, e.g. long bones or, arms and legs, femur and humerus.

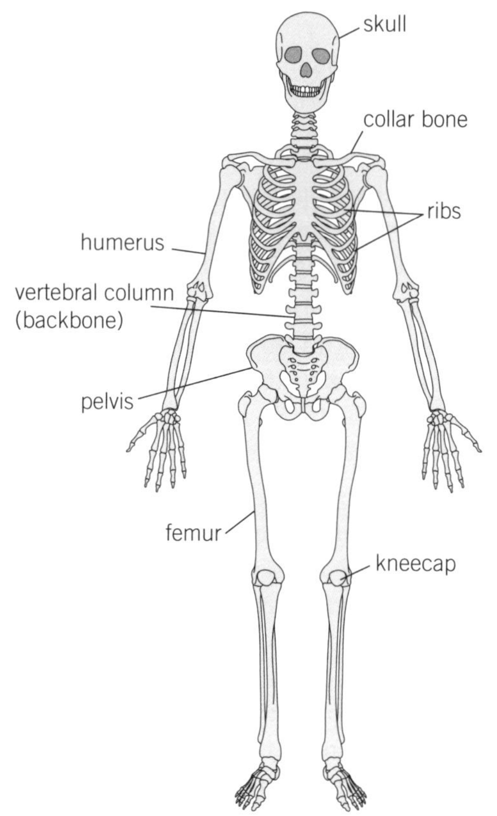

Extension

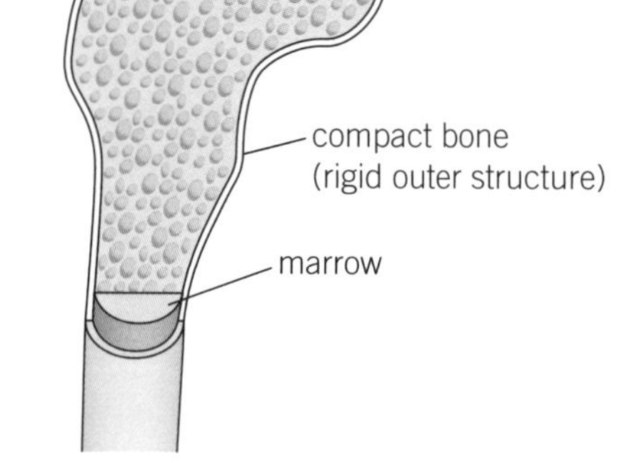

Labels/annotations: spongy bone – so it is not too heavy; compact bone – for strength; bone marrow – contains cells that make red blood cells and white blood cells.

6.10 Muscles and movement

1. A – a hinge joint, found in the knee, lets leg bend and straighten. B – a ball and socket joint, connects leg to hip, lets leg swing freely.
2. A3; B1; C2; D4
3. **a.** Karis **b.** Mikayla.
4.

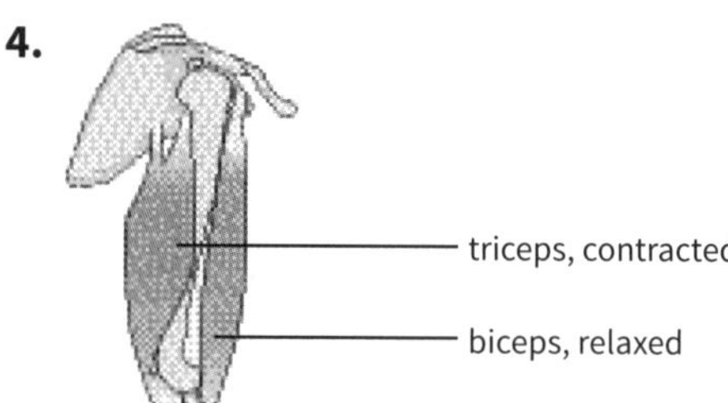

6.11 Health and inequality

You need to provide a piece of clear writing, including reference to scientific understanding in populations helping to solve issues linked to diet, the difference between deficiency diseases and starvation which are not usually a choice, and diseases linked to obesity which often do link to lifestyle choices, etc.

7.1 Ecosystems of the Earth

1. The missing words are: ecosystem; environment; ecosystem; biosphere; ecosystem; abiotic; biotic; soil/temperature; temperature/soil; light.
2. The food supply; the numbers of predators and prey; diseases and parasites.
3. A3; B5; C4; D1; E2

Extension

a. Biodiversity is a measure of the biological variety of an area.

b. Desert – low biodiversity; coral reef – high biodiversity; ocean – moderate biodiversity; temperate woodland – moderate biodiversity; tropical rain forest – high biodiversity.

7.2 Habitats within an ecosystem

1. **a.** Rainforest. **b.** rainforest **c.** desert **d.** desert **e.** rainforest **f.** desert.
2. **a.** Antarctic **b.** desert **c.** rainforest, **d.** Antarctic and desert, **e.** desert, **f.** rainforest, **g.** desert and Antarctic
3. Any three from: thick fur, small ears or large rounded body – to reduce heat loss; wide feet to avoid sinking in snow or to help the bear to swim; sharp claws to help catch its prey.

Extension

a. You need to give accurate descriptions of local habitat including temperature, water availability, changes through the year, etc.

b. You will need to have given three adaptations specific to the local habitat with a clear explanation of how they help the animal or plant survive.

7.3 The interdependence of organisms

1. A7; B6; C1; D5; E3; F2; G4
2. Population: the number of organisms of a particular type living in an area. Interdependent: species which affect each other within an area.
3. **a.** Any good example, e.g. grass ➔ caribou ➔ wolves

   b. Herbivore numbers increase as there are fewer carnivores to kill and eat them – producer numbers will fall as there are more herbivores eating them.

   c. Herbivore population might fall because the producer numbers fall because so many herbivores are eating them so there is not enough food to go around. OR Herbivore numbers might fall as carnivore numbers increase again and they start eating more herbivores.

Extension

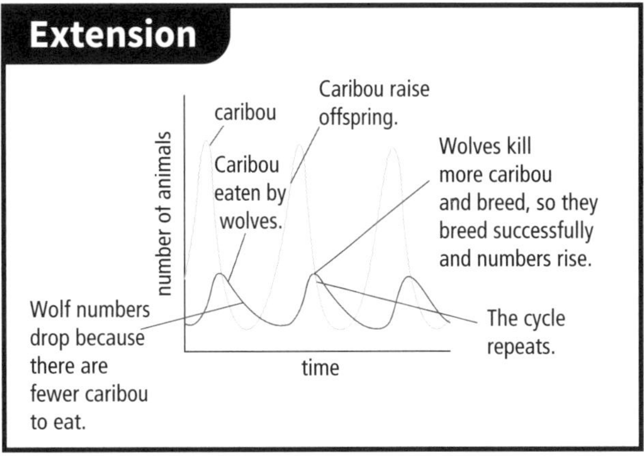

7.4 Pesticides and bioaccumulation

1. **a.** Water plants ➔ invertebrates ➔ small fish ➔ large fish ➔ fish-eating birds.

   b. The fish-eating birds eat the small and large fish, both of which accumulate high levels of pollutant. The concentration of pollutant increases up the food chain, and the fish-eating bird at the top of the chain accumulates the most.

   c. In high pollution areas, toxins accumulate through the food chain to a level which is high enough to kill the top predators, the fish-eating birds. When pollution levels are lower, the pesticides still accumulate in the top predators where they affect the thickness of the egg shells, making them thinner. Thinner egg shells are more likely to get broken as the baby birds are developing and fewer or no healthy chicks hatch and fledge successfully, so the adults fail to reproduce.

Extension

a.

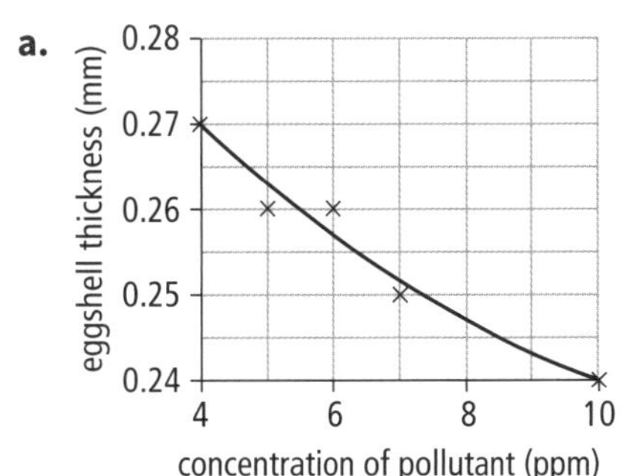

The graph shows that eggshell thickness decreases as the concentration of pollution increases.

b. The range of the results is too large to fit on normal scales – from 0.000 01 to 25.0.

7.5 Invasive species

1. A3; B4; C1; D2
2. The correct words are: invasive; ecosystem; damages; invasive; reproduce rapidly; new; no; population; breaks; an ecosystem.
3. **a.** Lionfish are a new, invasive species. They are carnivores and have no predators. They are quite big so they can feed on many native species. The graph shows their population grew very fast so they reproduced successfully. As a result, they have eaten most of the natural fish species in the areas where they are introduced. As the lionfish population grew, the prey fish populations fell steeply.
   b. The lionfish control problem involved finding that lionfish are good to eat; teaching people how to catch and eat them; using people as biological pest control.
   c. The graph shows that when lionfish were first introduced their numbers grew very rapidly and prey numbers fell very rapidly. The two were not synchronised. Once the lionfish control programme was introduced, lionfish numbers fell and prey numbers started to recover. Once lionfish numbers were controlled, a normal predator/prey cycle between the lionfish and the prey species can be seen.

7.6 Invasive species and ecosystems

1. The missing words are: food, disease, predators, extinct, invasive, quickly.
2. **a.**

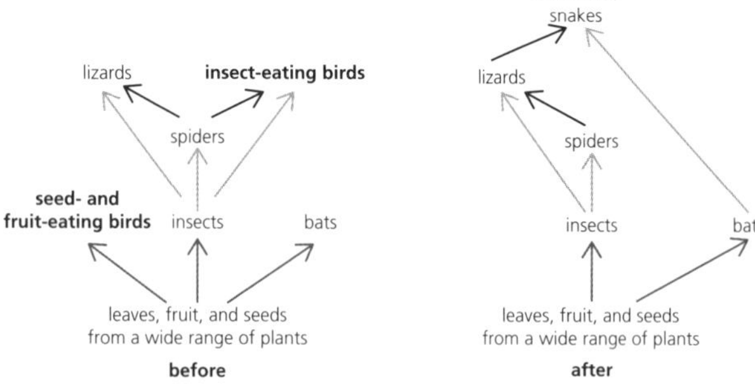

   b. Insect-eating birds – Down. Eaten by the brown tree snakes.

   Lizards – Down. Eaten by the brown tree snakes (but if you said 'up' or stays the same it is OK if you recognise that the lizards have less competition for their insect food once birds are gone so they may do better, at least for a time).

   Spiders – Up. Fewer insect-eating birds around to eat them.

   Bats – Down. Eaten by the brown tree snakes.

   Seed and fruit–eating birds – Down. Eaten by the brown tree snakes.

   c. If the insect population increases they could destroy the forest by eating all the leaves on the trees.

Extension

Marks for any logical prediction about the future of the forests on Guam, e.g. the loss of insect-eating birds will have the biggest effect in the short term because there will be more insects increasing the amount of damage caused by insects to trees; in the long term, the loss of seed- and fruit-eating birds will cause most damage because the existing trees will not be able to reproduce successfully.

7.7 Sampling your ecosystems

1. The missing words are: species; estimate; sampling; populations; quadrats; sampled; calculated.
2. **a.** A quadrat is an area used to study the number and distribution of organisms in an ecosystem.
   b. The sample areas must be chosen at random: divide the whole area into a grid and use a random number generator to give the coordinates of the sampling area to place the quadrat. Count the organisms in the sample, deciding whether or not to count organisms that touch the sides of the quadrat. Record the results. A minimum of three samples should be taken.
   c. Counting the thistles – in this case including ones that touch the sides: A= 5, B = 1, C = 6, D = 2.

   Mean for Tuns' field = $\frac{5+6}{2} = \frac{11}{2}$ = 5.5 thistles per 0.25 m^2 quadrat = 5.5 × 4 = 22 thistles per m^2 field.

   Mean for Jaz's field = $\frac{1+2}{2} = \frac{3}{2}$ = 1.5 thistles per 0.25 m^2 quadrat = 1.5 × 4 = 6 thistles per m^2 field.

   d. Based on these results, Jaz's claims seem to be true – his fields have far fewer thistles than Tun's so it seems likely they are spreading to his land from the field full of thistles.

   e. Only two quadrats were taken in each field – taking more random quadrats in each field would make the results more reliable and valid.

8.1 What is weather?

1. The missing words are: weather; atmosphere; air pressure; wind; humidity; temperature; clouds; precipitation; precipitation; snow/hail/sleet; hail/sleet/snow; sleet/hail/snow.

2. Weather is the mix of atmospheric events taking place in a particular place at a particular time.

3. ; thunderstorms; ; windy; cloudy.

Extension

Weather often varies and is very local because features like hills, mountains, and lakes change how the atmosphere behaves. This means that cloud formation, precipitation, etc., can vary over relatively small areas.

8.2 Climate and climate change

1. **a.** Climate describes the long-term weather patterns in an area.

   b. Weather and climate both involve factors such as amount of sun, wind, rainfall, humidity, and temperature.

   c. The main difference is the time scale. Climate is measured over many years. It tells us the patterns in the weather and therefore tells us the sort of weather to expect. Weather is short term and immediate – it is the weather conditions we experience at a particular time. Summarized: climate is what you expect; weather is what you get.

2. A4; B5; C6; D3; E1; F2

3. **a.** Any valid point such as fossils of tropical animals, e.g. trilobites found in countries that are now temperate.

   b. Any sensible point, e.g. temperature measured at the surface of the Earth is going up.

8.3 Climate changes past and present

1. **a.** The position of the Earth on its orbit round the Sun changes over thousands of years. When it is nearer the Sun, the climate is hotter/when it is further away from the Sun, the climate is cooler.

   b. Volcanoes and big meteorite strikes fill the atmosphere with dust and ash which stops sunlight reaching the surface of the Earth so it cools down.

   c. Sometimes the Sun is more active than others – when it is very active the Earth gets hotter/when it is less active the Earth cools down.

   d. Carbon dioxide traps heat energy from the Sun so, when levels in the atmosphere go up, the Earth gets warmer/when levels in the atmosphere go down the Earth cools.

2. **a.** Cycles of cooling and warming which have happened on Earth for thousands of years.

   b. An Ice Age is a period of time that lasts around 80–90 000 years when the surface of the Earth becomes very cold and much of the surface is covered with ice/oceans are frozen, etc. You should label one of the troughs on the graph where the temperature is low.

   c. Answers will vary depending on Ice Age chosen but will be around 200 ppm or less.

   d. You should label one of the peaks on the graph where the temperature is high.

   e. Answers will vary depending on temperature peak chosen but will be 280 ppm or more.

   f. It is useful because carbon dioxide levels in the atmosphere are rising rapidly – it helps scientists predict what will happen to global temperatures and also helps communicate to non-scientists what is happening.

Extension

Similar to historic patterns: carbon dioxide levels rising to over 280 ppm and accompanied by a rise in global temperatures.

Different to historic patterns: rise in carbon dioxide much faster than ever seen before and likely to go higher than seen before so we don't know what will happen or how living organisms will adapt.

8.4 Gathering evidence of climate change

1.

Measurements taken	What the evidence shows
past temperature readings from weather stations all over the world over many years to the present	the average temperature of the Earth's surface is rising over time
the thickness of tree rings on very old trees, showing how fast the trees grew each year	the temperature of the Earth has risen and fallen in the past – some years were much warmer than others
the concentration of carbon dioxide in atmospheric air trapped in layers of ice thousands of years ago	the concentration of carbon dioxide in the air/atmosphere has risen and fallen over time

2. a.

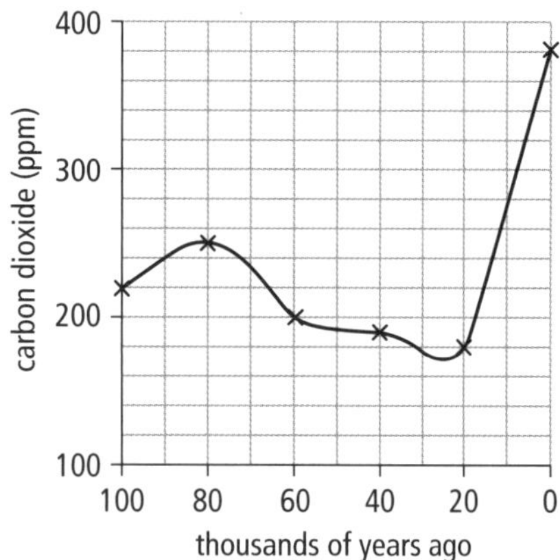

b. Carbon dioxide levels are 200 ppm higher than they were 20 000 years ago.

Extension

Scientists are worried because carbon dioxide levels are rising faster and higher than they have over the last 100 000 years. A rise in carbon dioxide levels is usually linked to a rise in surface temperature levels. High carbon dioxide could mean very high temperatures, significant changes in climate, melting ice, etc.

9.1 What do we know about plants?

1. **a.** A = roots, B = leaf/leaves; C = stem; D = flower.

   A: The roots anchor the plant in the ground and supply it with water and mineral nutrients from the soil.

   B: Leaves capture energy from the Sun using the green-coloured chlorophyll. They use this to make carbohydrates.

   C: Stems support the leaves, holding them out to capture the sunlight they need. They also support the flowers and fruit.

   D: Flowers are the reproductive structures of plants, and they form the seeds and fruit. They only appear at certain times in the life cycle of the plant.

2. Any three from: Plants are the producers for food chains and food webs all over the world. Plant crops are the source of most human food, both directly and as animal feed. Plants are part of the water cycle. Materials from plants are used for medicines, clothing, building, biofuels, and more.

Extension

Stems – swollen and fat to store a lot of water, green to capture sunlight to make food/photosynthesise.

Leaves – reduced to long sharp spines to stop animals eating them/reduced surface area reduces water loss from the plant.

9.2 Photosynthesis

1. **a.** Photosynthesis is the process by which plants make their own food using carbon dioxide from the air and water from the soil. They use energy from light. They produce glucose and oxygen.

   b.

$$\text{carbon dioxide} + \text{water} \xrightarrow[\text{chlorophyll}]{\text{light}} \text{glucose} + \text{oxygen}$$

2. a. A chloroplast is a structure found in many plant cells where the photosynthesis reactions take place.
   b. Chlorophyll.
   c. Not correct because not all plant cells contain chloroplasts. Only plant cells that are exposed to light contain chloroplasts and so are green. Structures such as the roots which are underground get no light so cannot photosynthesize and do not contain chloroplasts.
3. For aerobic respiration in the cells, making energy available for all the other reactions of life.

   To make starch as an energy store – glucose molecules are joined together to make large starch molecules which are stored in the leaves and in root and stem stores.

   To make all the other molecules needed in the plant such as cellulose for the cell walls.

9.3 Evidence of photosynthesis: testing for starch

1. The missing words are: variables; measured; changed; variable; repeat; reliable.
2. 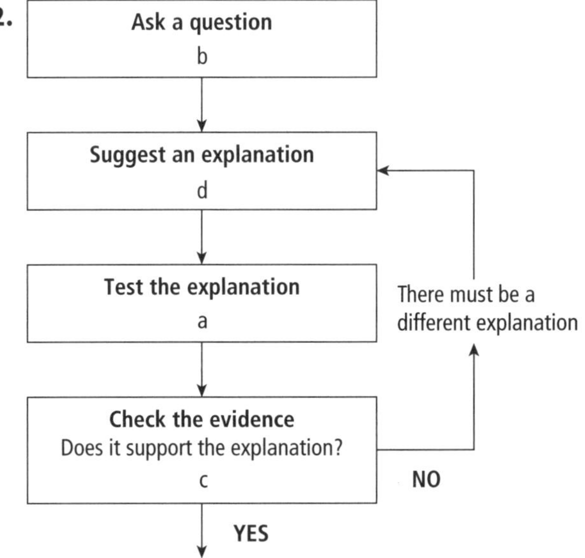

3. a. 1 Drop the leaf into boiling water to remove the waterproof outer layer, kill and break open cells. Turn off heat and remove leaf using forceps.

   2 Place leaf in test tube of ethanol and put test tube in hot water. As the ethanol boils, the green colour comes out of the leaf.

   3 Remove the stiff white leaf from the ethanol and dip it into the hot water to wash off the ethanol and soften it.

   4 Spread the leaf on a white tile and add a few drops of yellow-brown iodine solution to test for the presence of starch.

   b. Blue-black.

Extension

When leaves photosynthesise, they use carbon dioxide and water with energy from light to make glucose and oxygen. The glucose molecules are joined together to make starch which is used as an energy store in the leaves and other parts of plants. So the presence of starch indicates that photosynthesis has been taking place. When plants are kept in the dark they use up their starch stores.

9.4 Evidence of photosynthesis: oxygen bubbles

1. a. Prediction: light intensity will affect the rate of photosynthesis and the higher the light intensity, the more oxygen will be produced. Explanation: plants need light to photosynthesise, and the more light there is available, the more they can photosynthesise. Oxygen is produced as a waste product of photosynthesis and this is what Rahul is collecting and measuring, so the more light there is, the more gas he will collect.
   b. 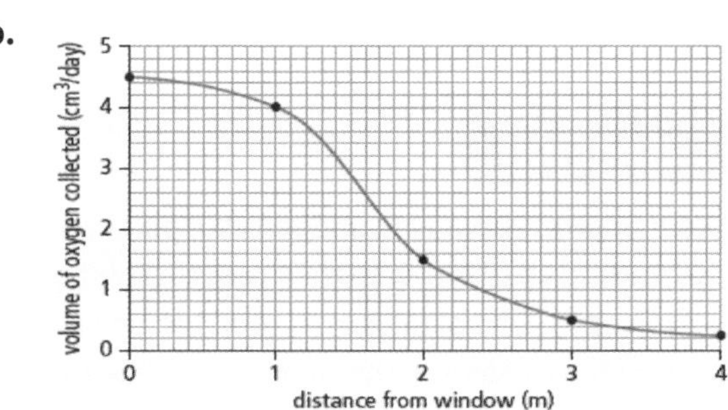

   c. Light intensity affects the volume of oxygen produced. The closer the pondweed is to the window, the higher the light intensity so more photosynthesis is carried out and more oxygen is produced.

Extension

a.

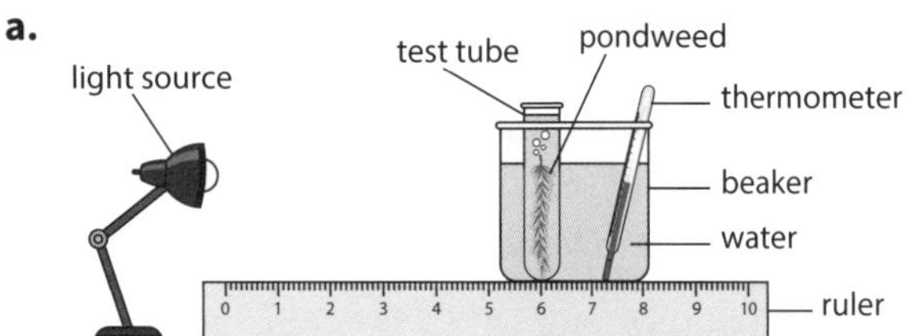

b. Prediction: rate of photosynthesis will increase as the carbon dioxide concentration of the water increases as plants need carbon dioxide as one of the raw materials for photosynthesis.

c. All the experimental beakers must be the same distance from the light source/window as light intensity also affects the rate of photosynthesis so they must all have the same light intensity.

9.5 The need for minerals

1. a. Nitrate deficiency causes poor growth and the older leaves turn yellow. Magnesium deficient leaves are pale or yellow.
   b. Nitrates are needed to make the proteins in plant cells. Proteins control the reactions in the cells and are part of the structure of the plant.
   c. Magnesium is needed to make chlorophyll which is used in photosynthesis.
2. The missing words are: minerals; soil; minerals; water; soil; plant; roots; legumes; nitrates; soil; root nodules; bacteria; nitrates; nitrogen; Legumes; nitrates.

Extension

Your answer should include sensible points such as: setting up plants with their roots in solutions of minerals. One plant is in pure water; one plant has all the minerals; other plants are in solutions missing just one mineral, e.g. no nitrates, no magnesium. Controlled variables, e.g. light and temperature. Measure/observe the height or biomass or number of leaves or colour of leaves – any sensible measure. Expect plants with no minerals will do badly, plants missing a single mineral will show signs of mineral deficiency over time.

9.6 The use of fertilisers

1. A2; B6; C5; D3; E1; F4
2. a. About 100 years.
   b. They work fast and are always available.
   c. They are expensive and do not improve soil structure.
3. a. In the early 20th century, a German chemist called Fritz Haber developed a way of making a compound called ammonia from the nitrogen in the air. Ammonia acts as a nitrate-rich fertiliser. Haber's laboratory method only made small amounts of ammonia so it was no use as a fertiliser but it was the start of the process.
   b. A chemist and engineer called Carl Bosch applied Haber's scientific discovery and developed a way to carry out Haber's reaction to make ammonia from air on a big scale, so it could be produced industrially. The Haber–Bosch process is still used to make millions of tonnes of nitrate fertiliser every year.

9.7 Water and mineral transport in plants

1. The missing words are: water; photosynthesis; absorb; vacuoles; cell; support; wilts
2. A Water evaporates out into the air through stomata on the leaves. B Water moves up through xylem vessels. C Water is taken in through root hair cells and moves into the roots.
3. A2; B1; C4; D5; E3

Extension

Petroleum jelly blocks the stomata on the underside of the leaf so water cannot be lost by evaporation. The plant with normal leaves loses water by transpiration from the leaves and so eventually it wilts as there is not enough water in the plant to maintain the cells. The plant with petroleum jelly does not wilt because it does not lose water by transpiration.

9.8 Xylem, phloem and plant pests

1. a. Tissue A = phloem and tissue B = xylem.

   b. Tissue A/phloem contains sugars in solution moving from the leaves around the plant.

      The aphid feeds on these sugars; tissue B/xylem contains water which has no food value for the aphid.

   c. Tissue A/phloem is a living tissue which uses energy to move sugars around the plant. Tissue B/xylem is dead tissue, made up of tubes that carry water from the roots to the leaves.

   d. Tomato plants free of aphids grow normally and well. Tomato plants infected by aphids may be smaller, and produce fewer tomatoes, because the aphids are using lots of the food made by the plant so it is not available for growth or to make fruit. The plants infected by aphids are also more likely to be diseased, which will also affect the yield of fruit. The aphids pierce the phloem and can introduce pathogens into the plant phloem. They can then be carried all around the plant, causing disease.

10.1 What is excretion?

1. The cells of your body carry out many reactions which result in waste products.

   Some of these waste products are toxic.

   The main poisonous waste product of aerobic respiration in your cells is carbon dioxide.

   The main poisonous waste from the breakdown of proteins in your body is urea.

   The removal of these substances from your body is called excretion.

2.

Lungs	Kidney
This is a pair of organs found in the chest, inside the ribcage.	This is a pair of organs found below the rib cage, against the back wall of the body.
They excrete carbon dioxide.	They excrete urea.
The carbon dioxide excreted is made in every cell during respiration.	The urea excreted is produced in the liver from the breakdown of excess proteins
The carbon dioxide is removed from the body when we breathe out.	The urea is removed from the body in the urine.

3. **Excretion**: The removal of waste products made in the cells of the body itself such as carbon dioxide and urea.

   Egestion: The removal of undigested material, which is taken into the body as food but cannot be broken down. It is passed out of the body as faeces.

10.2 The human excretory system

1. Compare your labels to Fig 10.2A in the Student Book. Kidney: filters the blood and removes urea, some water, and other substances your body doesn't need. Ureter: tube that carries urine from the kidney to the bladder. Bladder: stores urine until it is full. Urethra: tube that carries urine from the bladder to outside the body.
2. The missing words are: kidneys; blood; filter; blood; toxic; urea; blood cells; proteins; glucose; water; urine; urea.

Extension

It suggests that the kidneys, etc. are the only excretory organ in the body. Kidneys remove toxic urea but the lungs are also an excretory organ removing toxic carbon dioxide.

10.3 Who made the best model?

1. **a.** A model is used to help explain systems that are very big, very small, or difficult to see, or to predict what will happen in a system. There are physical models and computer models – the model shown in the diagram is a physical model.
   b. Any sensible answers, e.g. advantage – they can help us understand things that we can't see or that are very complicated by making them simpler/allowing us to see what is happening; disadvantage: they can give misconceptions because people take them very literally/they are always simplifications of the real thing.
   c. Advantage: any sensible answer, e.g. shows us something we can't see inside our bodies, shows us how the organs are arranged, shows us inside the kidney and the outside at the same time; disadvantage: any sensible answer, e.g. doesn't give us any idea how it works; oversimplifies it.
   d. You will be given credit for any model that makes sense and models the way the kidney works, e.g. simple filter set up. Advantages and disadvantages will depend on the model chosen but examples would be: helps understand the way the kidney works; big oversimplification can lead to misconceptions.

10.4 When kidneys go wrong

1. The missing words are: transplant; rejection; kidney; two; toxic waste; urea; urine; deceased; live.
2. A3; B4; C2; D1
3. **a.** T, **b.** F, **c.** F, **d.** T
   b. Organs can only be transplanted if the donor is a close tissue match to the patient.
   c. After a transplant, special medicines have to be taken for life to prevent rejection.

Extension

a. More organ transplants will be needed in future because the population is rising and people are living longer, which means that their organs are more likely to wear out.

b. A heart will be harder to grow than a bladder because it has a complex 3-D shape and contains several different sorts of tissue.

10.5 Kidneys work everywhere

1. **a., b. i. & c. i.**

Species	Mean daily urine output (cm^3/kg)
horse	10
cattle	31
sheep or goat	25
cat	30
b. i. kangaroo rat	0.5
c. i. beaver	50

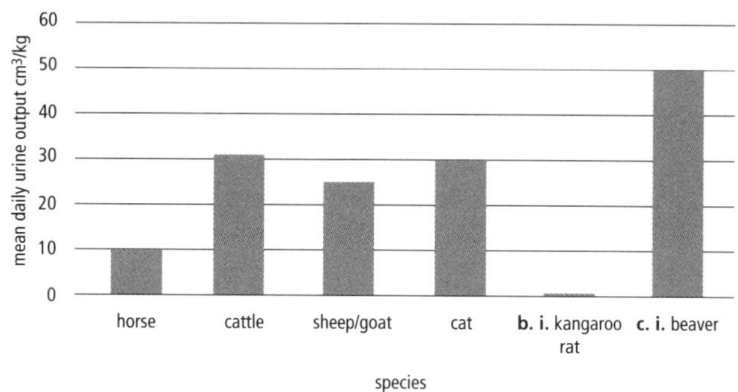

b. ii. Kangaroo rats live in very dry environments with very little water so their kidneys are very efficient and they produce very small amounts of very concentrated urine.

c. ii. Beavers live in fresh water so they have plenty to drink and do not use energy to concentrate their urine – their kidneys produce large amounts of very dilute urine.

11.1 Reproduction: a characteristic of life

1. A3; B4; C1; D2
2. The missing words are: two; asexual; offspring; sexual; two; offspring; DNA; similar; different.
3.

Species	Number of chromosomes in body cells	Number of chromosomes in gametes
human	46	**23**
elephant	**56**	28
coconut palm	32	**16**
boa constrictor	**36**	18
torch ginger	48	**24**
tortoise	**52**	26

Extension

a. The plants produced asexually will have exactly the same DNA/genetic material as the parent plant. The plants produced sexually will have different combinations of DNA/genetic material to the parent plant.

b. The asexual offspring are produced by the parent plant cells dividing in two and so they are identical. In sexual reproduction the parent plant forms gametes which have half the number of chromosomes of the parent plant. Gametes from different plants join to form the seed that grows into the new plant, so it is similar to but different from its parent.

11.2 Fertilisation: new life begins

1. The missing words are: egg; ovary; sperm; egg cell; sperm; nucleus; fertilisation; DNA; chromosomes.

2. **a.** A gamete is a special reproductive cell. A gamete differs from a normal body cell because it has only half the chromosomes of a normal cell and it is specialised for its functions in reproduction.

b.

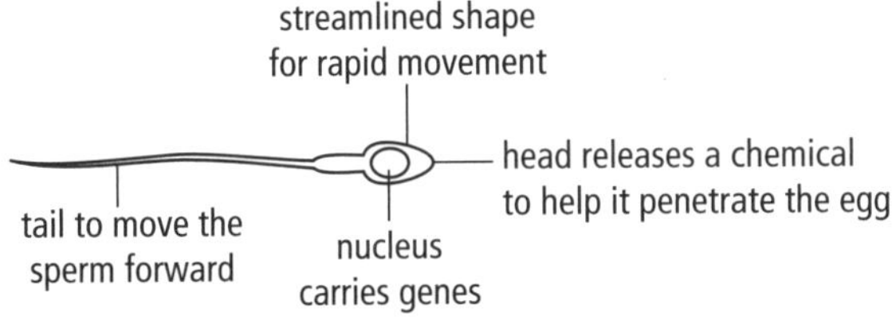

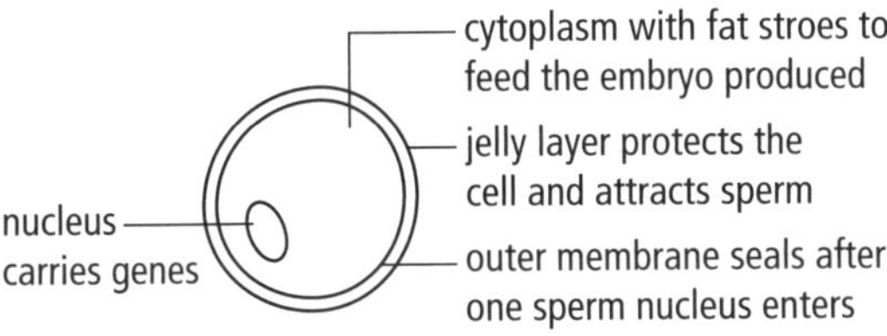

3. First diagram: sperm nucleus enters the egg cell.; Second diagram: sperm and egg nuclei fuse to create a new cell/life.; Third diagram: fertilised egg begins to divide.

Extension

The genes that make up the chromosomes control the features of the developing baby as it grows. The chromosomes inherited will also affect whether the new individual is a boy or a girl.

11.3 Boy or girl? Sex inheritance in humans

1. **1. a.** Both. **b.** Both. **c.** Chromosomes **d.** Chromosomes. **e.** Genes. **f.** Genes. **g.** Chromosomes.

2. **a.**

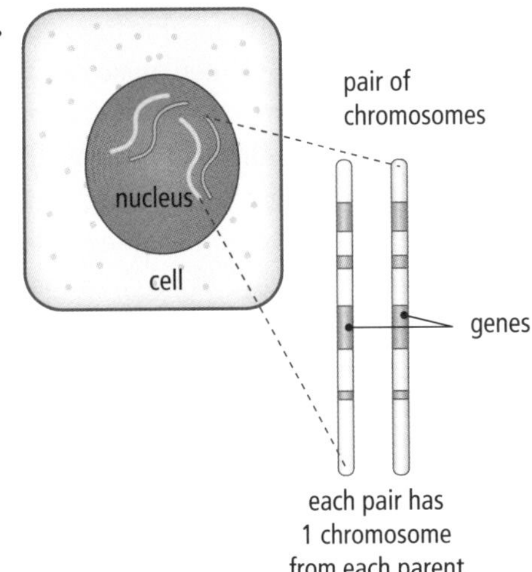

b. The nucleus in the diagram has two pairs of chromosomes, but real human cells have 23 pairs.

The chromosomes in the diagram each contain four different genes, but real human chromosomes can have thousands on genes.

c. In humans, 22 out of the 23 pairs of chromosomes are always identical to each other. The sex chromosomes may be the same or different The sex of the individual is determined by the sex chromosomes they inherit. Females have two X chromosomes but males have an X chromosome and a Y chromosome.

3. 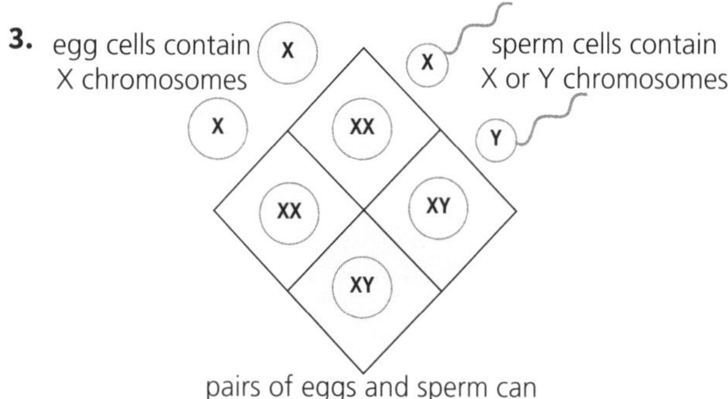

You inherit one of each pair of chromosomes from your mother and one from your father. Women's cells are XX, so all eggs contain an X chromosome. You will always get an X chromosome from your mother. Men's cells are XY, so men make two types of sperm. Half of them carry an X chromosome, and half of them carry a Y chromosome. All of the sperm have an equal chance of fertilising the egg. If a sperm containing an X chromosome fertilises the egg, the new cell will have two X chromosomes (XX) so the child will be a girl. If a sperm carrying a Y chromosome fertilises the egg, the new cell will have an X and a Y chromosome (XY) so the child will be a boy. The chances of a baby boy or a baby girl are always 1 in 2, or 50%.

11.4 Variation between individuals

1. **a.** T **b.** T **c.** F **d.** T **e.** T **f.** F **g.** F

   c. Identical twins always have identical genes.

   g. Cells specialise by switching on **different** genes.

   f. Every cell in your body except the gametes you produce contain identical genes.

2. **a.** 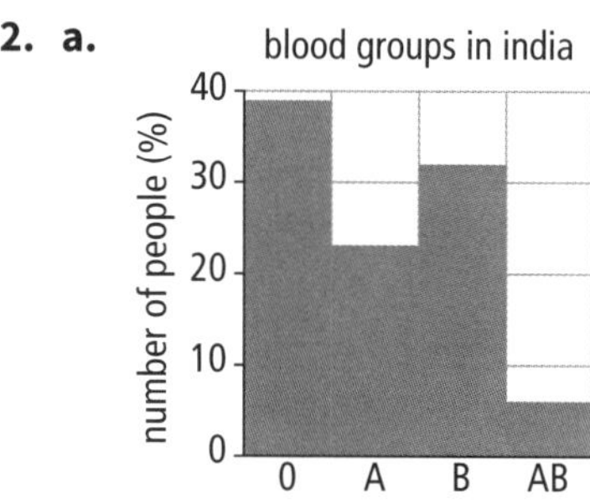

   b. Inherited variation is determined by the genes you inherit and nothing else, so nothing can change these characteristics. Each individual inherits their blood group and it will remain the same throughout life.

   c. Height is affected both by the genes you inherit and the environment you live in. The average height of students is increasing in many parts of the world because they eat more nutritious food and a bigger percentage reach the maximum height their genes can produce.

Extension

a. Identical twins have the same genes. If they are separated at birth, they will grow up in different environments. Any differences between them must be caused by these environmental differences.

b. If inherited variation is more important than environmental variation in determining behaviours, we would expect identical twins separated at birth to be very similar in their behaviours – almost as similar as identical twins who grew up together.

11.5 The development of a fetus

1. The correct order is: G, D, A, F, E, B, C.

2. **a.** 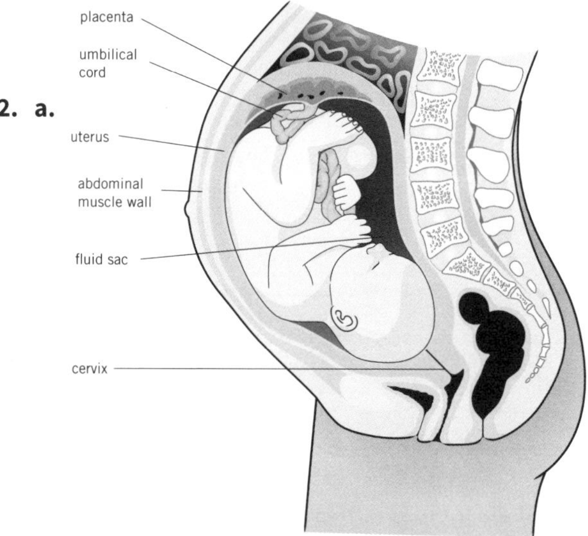

   b. The umbilical cord joins the fetus to the placenta, carrying blood loaded with waste products from the fetus to the placenta and bringing blood loaded with food and oxygen back to the fetus.

   c. The fluid in the fluid sac supports the baby as it grows, allowing it to move freely. It acts as a shock absorber and so protects the fetus from and bumps or knocks to the mother's body.

d. The placenta allows substances to move between the blood of the mother and the blood of the fetus. The blood of the mother and the blood of the fetus flow very close together but they do not mix. Dissolved food molecules and oxygen move by diffusion from the mother's blood to the fetus. Waste carbon dioxide and urea pass by diffusion from the fetus to the mother, to be removed. The placenta also acts as a protective barrier. It stops most infections and harmful substances from reaching the baby.

11.6 Health of the mother, health of the child

1. a.

mother's blood | blood of fetus

A

oxygen and glucose

B

corbon dioxide and urea

b. Some medicines which are safe for children and adults to take can damage a developing fetus. Some substances cross the placenta and get from the blood of the mother to the blood of the fetus. If a mother is ill and needs medicine during her pregnancy, her doctor must take care to only prescribe drugs which either do not harm the fetus or do not cross the placenta.

2. a. It gets nutrients from the mother as they cross the placenta by diffusion down a concentration gradient, moving from the blood of the mother to the blood of the fetus.

b. Pregnant women need iron in their diet to make the red blood cells/haemoglobin they need to carry enough oxygen for themselves and their developing fetus.

The developing fetus has to make red blood cells/haemoglobin – it needs iron to do this and its mother is the only place it can get the iron from, so she must eat plenty in her diet.

c. Calcium is needed to form strong teeth and bones. The fetus is growing teeth and bones so it needs calcium – it relies on the calcium it gets from its mother so it is important that the mother eats a high calcium diet to pass the mineral to the fetus through the placenta.

d. Any sensible choice based on scientific understanding, e.g. protein for the growth of cells and muscles in the fetus, vitamin D so the fetus can use calcium to build strong bones and teeth, etc.

Extension

Drugs can cross the placenta and may affect the fetus in a number of ways. Some reduce the growth of the baby, e.g. caffeine and this increases the risk of various conditions – see Fig 11.6 D in Student Book. Some drugs harm the fetus and stop it developing properly. Fetuses may become addicted to drugs such as heroin and cocaine before they are born.

11.7 Understanding science, saving lives

1. a.

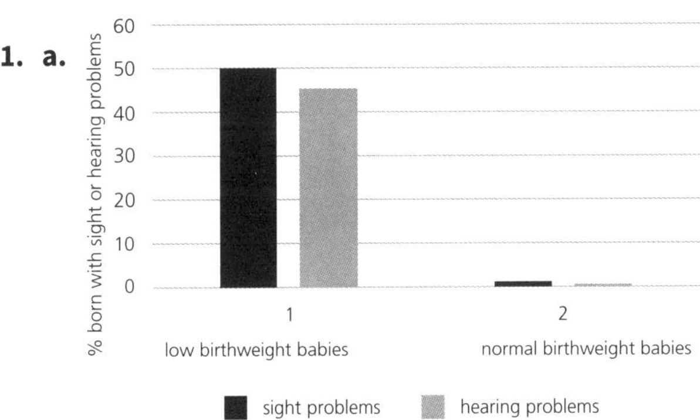

b. The data show that babies who are born weighing less that 1 kg have a much higher risk of having sight and hearing difficulties. Most of the weight that the fetus gains before birth is gained in the last three months of pregnancy so it is very important for women to have plenty of food during those last three months to ensure their baby is a healthy weight when it is born and so reduce the risk of sight and hearing problems.

2. Your answer should show you understand that the science behind low birth weights can be used to help reduce the problems of babies born with sight and hearing problems as a result, e.g. families/communities can prioritise food for pregnant women, women try to eat healthily, avoid substances that reduce birth weight, etc. BUT there are some situations where there is no scientific answer and so people cannot do anything to reduce the risks for themselves or their baby.

11.8 Smoking and pregnancy: the evidence

1. **a.** Low birthweight babies ÷ total babies born to smokers = 30 ÷ 74 = 40.5% of smokers' babies are low birthweight.

   Low birthweight babies/total babies born to non-smokers = 29 ÷ 115 = 25.2% of non-smokers' babies are low birthweight.

   b.

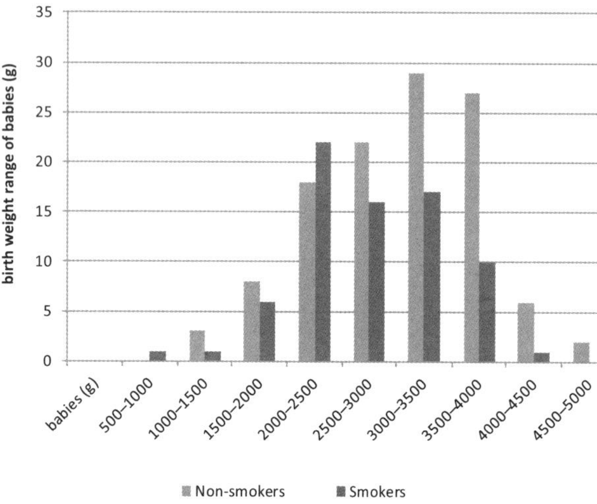

   c. Relatively small sample of 189.

   Don't know what their background/education was.

   Pattern very similar to data in student book – smoking in pregnancy increases the risk of a low birth weight baby and low birth weight babies do not develop as well as normal weight babies.

   d. Cigarette smoking decreases the amount of oxygen in the blood because it produces carbon monoxide which is carried in the blood. The fetus of a mother who smokes does not get enough oxygen and so cannot grow as well.

12.1 The carbon cycle

1. **Photosynthesis**: takes place in the green parts of plants and in algae. It uses energy from light, trapped by chlorophyll in the chloroplasts. Carbon dioxide is taken in from the air and combined with water to produce glucose and oxygen. This in turn is used to build the biomass of plants. Photosynthesis is the first stage of most food chains. **Feeding**: animals feed. They eat plants or other animals. They take in the biomass of plants, break it down, and build it back up into animal biomass. Carbon is passed from one organism to another in this way. Some is released into the environment through respiration at the same time. **Respiration**: takes place in all living organisms. Glucose molecules are broken down using oxygen, producing carbon dioxide (CO_2) and water. This releases energy to be used by cells. **Decomposition:** the process carried out by the decomposers. Microorganisms such as bacteria and fungi feed on the waste materials produced by animals, the dead leaves produced by plants, and the dead bodies of animals and plants. They break them down, releasing carbon dioxide back into the atmosphere. **Combustion:** the scientific word for burning. When a substance burns, it reacts with oxygen in the air and the energy transferred heats the surroundings and produces light. Many fuels, from wood to fossil fuels like coal, oil, and gas, are carbon compounds. When these fuels burn, they produce carbon dioxide and water which are released into the atmosphere.

Extension

The balance of the carbon cycle is being disturbed by human activities such as burning fossil fuels and cutting down trees. The effect is global warming and climate change. Unless people understand the science and how the carbon cycle works, it is hard to show them how we can help to restore the balance in the carbon cycle.

12.2 People and the carbon cycle

1. The missing words are: Earth; water/carbon/oxygen; carbon/oxygen/water; oxygen/water/carbon; temperature; organisms; live; greenhouse gases; carbon dioxide; energy; Sun; warm.

2. **a.** A rise in the concentration of carbon dioxide in the atmosphere causes an increase in the temperature at the surface of the Earth. This is because carbon dioxide is a greenhouse

gas that traps some of the Sun's energy in the atmosphere. As the level of carbon dioxide increases, more energy is trapped in the atmosphere and so the temperature at the surface of the Earth rises.

b. Any three sensible points with a relevant explanation of the effect, e.g. increasing transport by cars, lorries, planes, etc., all of which burn fossil fuels in their engines. The combustion produces carbon dioxide so the levels increase as transport increases. More people = more electricity generated. Much electricity generation is based on burning fossil fuels which again increases the levels of carbon dioxide in the atmosphere. Deforestation involves cutting down trees – trees take up carbon dioxide in photosynthesis so deforestation reduces the removal of carbon dioxide and so increases the levels of carbon dioxide in the atmosphere.

Extension

Your answer should show your awareness of different carbon dioxide sources in richer countries with more vehicles/flying/electricity generation and use; the need to balance development of poorer countries with increases in carbon dioxide production and ways of reducing the carbon dioxide production of richer countries.

12.3 Historical impacts of climate change

1. A4, B1, C5, D2, E3

2. **A** Huge volcanoes erupting, filling the atmosphere with ash and blocking the sunlight. **B** Giant meteorites striking Earth, causing dust that blocked out the Sun. **C** Changes in the orbit of the Earth so it was further from the Sun. **D** Changes in the activity of the Sun.

3.

Observed change	Explanation
sea levels rising	Global warming melts land ice, and mean sea level rises. Rises in sea level increase coastal erosion and changes in ocean currents. This impacts the movements of the tiny organisms supporting all the ocean food chains.
flooding	Global warming and climate change are linked to increases in very heavy rainfall which lasts a long time and causes this problem.
drought	Climate change can lead to extreme heat and a lack of rain with crops failing and lakes drying. Many people can no longer grow enough to eat. The whole ecosystem is affected. Droughts increase the risk of wild fires which destroy whole ecosystems.
extreme weather events	Global warming changing the airflows in the atmosphere, affecting the climate everywhere and causing more extreme weather events such as hurricanes, tornados, blizzards, heavy rain, and very high temperatures.

12.4 Predicting the future

1. **a.** Doubling the amount of carbon dioxide in the atmosphere would make it 5 °C warmer at the surface of the Earth.

   b. They show atmospheric carbon dioxide levels increasing and the surface temperature increasing in the same pattern. They are not quite increasing at the rate Arrhenius predicted but they are close.

2. Any 3 from:
   - As oceans get warmer, and glaciers and land ice melts, some scientists predict a rise in sea levels of 26–82 cm by the end of the 21st century. Low lying countries like Tuvalu, and areas of many other countries, will disappear beneath the sea.
   - As temperatures rise and rainfall patterns change, huge areas will become infertile. Plants and animal species will die out if there is no water to support life. People will starve.
   - Many parts of the world will get more rainfall and snowfall than ever before, often as extreme

storms or blizzards (snow storms). This will lead to floods, washing away fertile soil and causing the loss of homes and lives.

- Some countries will get hotter, so new pests such as mosquitoes, ticks, and crop pests will survive, impacting health and agriculture. Some will get colder. Ecosystems are often very temperature dependent, so plant life, migration patterns of animals, and the species you see around every day will change in future. Many species will become extinct.
- If climate change continues, extreme weather events such as hurricanes and storms will happen more often and affect more of our lives. A single storm can cause huge damage to infrastructure such as power supplies, buildings, roads, and railways.

Extension

Range of temperatures because of the unpredictability of: people's behaviour, events such as volcanoes, legislation to control carbon emissions, etc.

Any sensible points including human behaviour – changing how we generate electricity, burning less fossil fuel, whether we continue rate of deforestation of slow it down, etc.

12.5 Evaluating the evidence for climate change

1. a.

Measurements taken	What the evidence shows
past temperature readings from weather stations all over the world	Earth's average temperature is rising.
the thickness of tree rings	Because the rings are thicker when the weather is warm and smaller when it is cold, this shows the Earth's temperature has risen and fallen in the past.
the amount of carbon dioxide trapped in layers of ice	This shows how much carbon dioxide was in the atmosphere so the amount of carbon dioxide in the atmosphere has risen and fallen in the past.

b. It reveals patterns in the climate of the Earth over time. It shows links between the rise and fall of carbon dioxide in the atmosphere and the temperature at the surface of the Earth. Any other sensible point.

2. a. The levels of carbon dioxide in the air from thousands of years ago trapped in bubbles in the ice.

b. Scientists record the changes in carbon dioxide levels in the atmosphere over thousands of years and compare it to levels of carbon dioxide recorded in the atmosphere now.

c. Over the last two centuries the level of carbon dioxide in the atmosphere has been rising steadily. Many scientists worry because the evidence tells them that when carbon dioxide levels rise the temperature at the surface of the Earth increases, and carbon dioxide levels are rising faster and have reached a higher level than they have ever been in the last 100 000 years.

13.1 Variation in animals and plants

1. The missing words are: inherited; genes; offspring; environmental; environment; characteristics; combination; genetic/environmental; environmental/genetic.

2. a. They have many different characteristics because there is a lot of genetic difference between them.

b. Organisms of the same species have lots of shared genetic characteristics which is why they look similar and can breed together successfully.

c. Organisms of the same species are not genetically identical so they have some different characteristics. They also grow up in different environments in different conditions which means they have differences resulting from their environments.

3. Reasons for the variation: both leg length and body mass are affected by genes so these families of lion cubs will be different as a result of their different genes. However, body mass in particular is also affected by environment. Perhaps one mother lion brings back more food for her cubs than the other. The best fed cubs will be able to fulfil their genetic potential and gain body mass and their legs will grow as much as possible. Cubs with less food will not be able to grow as well or gain as much body mass.

13.2 Natural selection in action

1. From top of the table downwards: 3, 5, 4, 1, 2.
2. **a.** They all have a similar number of bones arranged in the same order but their sizes and shapes are different in each species.
   b. The bat finger bones are extended so they hold out its wings to give a large surface area to help it fly. All the bones are very thin to reduce their mass, which also makes flight easier.
   c. The arm bones are very short and wide for strength to push against the water and the central finger bones are extended to give the fin a streamlined shape.

Extension

a. Some bacteria in a population may have some resistance to an antibiotic. These are the bacteria that will survive and pass on those successful genes to their offspring. The most resistant offspring will survive and reproduce – so gradually, through natural selection, the whole population becomes resistant to the antibiotic.

b. If a population of bacteria become resistant to an antibiotic through natural selection, it makes it difficult to cure the disease. If bacteria become resistant to several different antibiotics through natural selection, a disease may become untreatable and people may die.

13.3 Environmental change and natural selection

1. The missing words are: species; characteristics; survive; offspring; characteristics; generation; common; population; natural selection.
2. **a.** White fur provides better camouflage against ice and snow which makes it easier for white bears to catch their prey. It also means they are less likely to be seen and killed by other brown bears when they are young.
   b. Bears with white fur living in cold, icy places are more successful hunters than brown bears so they are more likely to survive and reproduce.

   The genes that made them successful are passed to the next generation. Over many generations, these genes become more common and the whole population eventually has white hair.

   c. If all the environments where bears live had become and remain icy and snowy, probably only polar bears would exist. Brown bears would struggle to hunt successfully in a white environment because they show up. However, only the Arctic has a white, icy environment. In other areas, white bears would show up much more than brown bears so they would not hunt successfully. The different species of bears are adapted to different environments, and as long as those environments are there, the different bear species will survive.
   d. **Why:** global warming and climate change means the world is warming up and the ice/snow in the Arctic may disappear. **How:** if there is no ice and snow, white polar bears would show up in their environment and be unable to hunt effectively. They would be outcompeted by brown bears and die out /darker cubs would be more likely to survive and white coats would die out by natural selection.

13.4 Extinction!

1. The missing words are: plant; tropical; biomass; habitats; biodiversity; habitats; destroyed; extinct.
2. **a.**

Vertebrate group	Threatened species (% of total in 2000)	Threatened species (% of total in 2020)
fish	3	20
amphibians	3	41
reptiles	4	34
birds	12	14
mammals	24	26

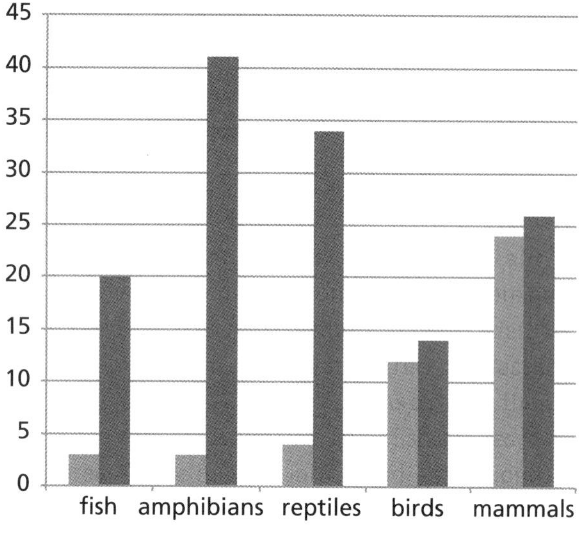

b. Amphibians and reptiles have seen the greatest increase in threatened extinctions, while birds and mammals have seen the lowest percentage increase – although they have maintained high levels of threat throughout the 20 year period.

c. Human impacts are the main causes – look for three clear reasons; e.g. increased habitat destruction; increased pollution; increased hunting/fishing; changes in land use; draining of wetlands, etc.

Extension

Plant seeds can be collected and stored for many years until habitats are restored. Animals cannot be put in cold storage in this way so conservation has to be much more immediate. They have to be conserved and so do the places where they live, otherwise they are lost forever.

13.5 Investigating the peppered moth: past and present

1. A5; B4; C2; D3; E1

2. a. The data describe the changes in the percentage of different coloured *Biston betularia* moths captured between 1800 and the early 2000s in the UK.

   They show that, from the mid-1800s, the numbers of light-coloured moths caught dropped dramatically from over 95% to less than 5% of those caught, and the percentage of dark moths caught increased over the same timescale from less than 2% to 95% or more.

   In the 1970s, this trend was reversed and, by the early 2000s, the numbers of pale moths caught climbed back up to almost 100% and the numbers of dark moths caught fell to almost 0%.

   b. Initial 1800s natural populations were largely pale moths camouflaged on light tree trunks; occasional dark forms could be easily seen and eaten. After the Industrial Revolution, many trees and buildings were blackened by soot and smoke. Pale moths could now be easily seen. Darker moths were less visible so they survived to reproduce. Over about 50 years, the colour proportions swapped over so dark became the majority and there were fewer pale moths through natural selection. In the 1970s, laws were made to make the air cleaner. Without the pollution the trees became their natural pale colour again. Now dark moths were easy for birds to see and eat. Pale moths were more likely to survive and reproduce and, by natural selection, the balance of colour in the population returned to mainly pale.

13.6 What can we do?

1. a. Biggest increase: Middle East and Pakistan. Lowest increase: Europe and United States.

   b. i. A country may have greatly increased its production of electricity using hydropower OR it may increase from no generation of electricity using water to a tiny amount of generation which would appear as a big percentage increase.

   ii. A country may already produce most of its electricity using hydroelectric power and increasing it more may only give a small percentage increase overall OR the country may have made little or no increase in using hydropower to generate electricity.

   c. i. Reduces the burning of fossil fuels to generate electricity. This reduces the amount of carbon dioxide released into the atmosphere, in turn reducing the greenhouse effect and reducing global warming and climate change.

   ii. Either: uses water that may be needed elsewhere for drinking or farming causing famine. Or: threatens fish biodiversity by preventing them moving up and down rivers to breed.

Glossary

Abiotic Describes something that is non-living.

Adaptation A physical characteristic or behaviour that makes animals or plants particularly suited to their environment.

Aerobic respiration The controlled release of energy that takes place in the mitochondria of the cells, using oxygen.

Amino acids Small molecules that are the building blocks of proteins.

Anaerobic respiration The controlled release of energy in cells without using oxygen. It produces much less available energy than aerobic respiration.

Artificial fertilisers Fertilisers that are compounds made by industrial processes.

Asexual reproduction Reproduction involving only one parent, that results in offspring that are identical to their parent.

Bacteria (sing. Bacterium) Some of the smallest living organisms, between 0.2 and 2.0 μm in diameter. They are all single celled and reproduce by splitting in two.

Bioaccumulation The increase in the concentration of a substance along a food chain, often a toxin.

Biodiversity The number and variety of living organisms.

Biomass The material living organisms are made of.

Biotic Describes something that is living or is part of a living organism.

Calcium Mineral needed for making strong, healthy bones and teeth.

Cancer A group of diseases where the cells divide uncontrollably.

Capillaries Very small blood vessels.

Carbon cycle A series of processes that move carbon between living organisms and the physical environment.

Carbon dioxide (CO_2) A gas produced during aerobic respiration in cells and used by plants for photosynthesis. It is a greenhouse gas.

Carbon monoxide (CO) A poisonous gas that reduces the amount of oxygen the red blood cells can carry.

Cell(s) The single units or building blocks of living things.

Characteristics Features of living organisms that can be observed and measured.

Chlorophyll The green coloured pigment which traps light energy for photosynthesis in the chloroplasts of plant cells.

Cilia Tiny hair-like structures found on some cells (ciliated cells) that beat to cause movement.

Classification Sorting organisms into groups based on their similarities and differences.

Climate The long term weather patterns in an area.

Combustion The scientific word for burning.

Concentration gradient The difference in concentration between an area with a higher concentration of a substance and an area with a lower concentration of the same substance.

Controlled variable Variable that is not changed during an investigation.

Deficiency disease A disease which results if a diet is lacking a certain element long term.

Embryo The early stages of development after fertilisation.

Estimate Approximate calculation or judgement of an answer based on knowledge and understanding.

Evaluate Work out what is good and what is not so good.

Extinct When no more individuals of a species remain alive, either locally or globally.

Fertilisation The point at which the nucleus of the egg and sperm fuse to make a new cell with a full set of chromosomes.

Fetus The developing baby after the initial stages of pregnancy.

Fibre An important part of the diet made up of large molecules the body can't digest that helps everything move steadily through the digestive system.

Flowering plant (seed plant) Plants that reproduce using flowers and making seeds.

Fluid sac The fluid filled structure that surrounds a developing fetus to support and protect it.

Fossil The preserved remains of ancient organisms.

Function Job or role in a cell or organism.

Gametes Special reproductive cells that contain half the number of chromosomes as the normal body cells of the parent.

Germ theory of disease The idea that infectious diseases are caused by microorganisms (germs) passed from one organism to another.

Glacier A huge frozen river of ice.

Glycogen A short term energy store in mammalian muscles and liver.

Greenhouse effect The effect of greenhouse gases warming the surface of the Earth.

Greenhouse gases Gases that reflect heat back to the surface of the Earth, maintaining a temperature suitable for life.

Growth Getting bigger by gaining more cells or more biomass.

Habitat The area where an organism lives – its home.

Identical twins Twins formed from a single fertilised egg, who are genetically identical.

Inherited variation Characteristics determined solely by your genes.

Introduced species (non-native species) Species that are brought into a country or an ecosystem by human activity.

Invasive species A species introduced into an ecosystem by human activity, which then damages the ecosystem.

Iron Mineral needed to make the haemoglobin that carries oxygen around the body.

Joint The structures where two bones meet.

Light intensity Measure of the amount of light falling on an object.

Lipid Fats and oils, needed to supply energy, used as an energy store in the body and needed for cell membranes. Butter, cheese and meat are fat-rich foods.

Micron (μm) 1/1 000 000 of a metre

Mineral deficiency A lack of a mineral compound needed for health and/or growth.

Mineral salts/minerals Dissolved compounds needed by plant and animal cells to grow and remain healthy.

Misconception A wrong or inaccurate idea.

Mould A type of fungus that feeds on many different materials. Some moulds produce antibiotics like penicillin.

Multicellular Made up of many cells.

Myoglobin A substance similar to haemoglobin that stores oxygen in the muscles.

Native species Species that are a natural part of an ecosystem.

Natural fertilisers Fertilisers made from natural materials that have been decomposed eg compost, manure.

Natural selection The process by which organisms with the characteristics best adapted to their environment live and reproduce, passing on the useful characteristics to their offspring.

Net Overall.

Non-flowering plant Plant that reproduces using spores. They do not produce flowers.

Obese Very overweight.

Palisade cells Specialised plant cells found near the top surface of leaves which contain many chloroplasts for photosynthesis.

Peer review The process by which scientists check the work of other scientists before it is published, to make sure it is reliable and can be repeated.

Phloem Specialised living plant tissue that transports dissolved food from the leaves around the plant.

Photosynthesis The process by which plants make their own food from carbon dioxide and water, using light energy.

Placenta The structure that allows substances to pass between the mother and the fetus throughout pregnancy, providing the fetus with food and oxygen and removing waste products produced by the fetus.

Plasma The yellow, liquid part of the blood that carries cells and many dissolved substances around the body.

Platelets Tiny pieces of cells that help the blood clot.

Pregnancy/Gestation period The time it takes a baby to develop from a fertilised egg to birth.

Protein Substance needed to build new tissues, replace old tissues and repair damaged tissues; found in meat, fish, eggs, beans and nuts.

Quadrat An area used to study the numbers and distribution of organisms in the field.

Red blood cells Tiny, flexible cells with a biconcave shape and no nucleus when they are mature. They are filled with haemoglobin and carry oxygen around the body.

Reproduction The process by which living organisms make more of themselves (produce offspring).

Sexual reproduction Reproduction involving two gametes and usually two parents. It results in offspring that are similar to but different from both parents.

Single-celled organisms Organisms made up of only one cell that carries out all the functions of life.

Solvent A substance that other substances will dissolve in.

Specialised cells Cells adapted to carry out particular functions.

Species A group of organisms with similar characteristics that reproduce to give fertile offspring.

Specimen Object being observed e.g. the organism or part of an organism looked at through a microscope.

Spores The reproductive cells of non-flowering plants and fungi.

Starch A carbohydrate made of many sugar molecules, used as an energy store in plants.

Surface area The outside area.

Tar The sticky black substance in tobacco smoke that causes cancer.

Toxic Poisonous.

Toxin Poison.

Transpiration The evaporation of water from the surface of leaves.

Uterus The organ where a developing baby grows during pregnancy.

Vaccine Material given to develop an immune response to protect against infection by a specific pathogen e.g. polio, COVID-19.

Vitamins Substances needed in tiny amounts in the diet to help chemical reactions take place in cells.

Virus Extremely tiny structures about 0.1–0.01 μm, made of genetic material inside a protein coat. They can only reproduce inside other living cells.

White blood cells Large blood cells with a nucleus that protect the body against pathogens by producing antibodies or by digesting microorganisms.

Xylem Specialised dead tissue that transports water and dissolved mineral salts from the soil up to the leaves and flowers of the plant.

Yield A measure of the amount of a particular product produced, particularly applied to crops.